FOOD SCIENCE AND TECHNOLOGY SERIES

LIQUID CHROMATOGRAPHY FOR THE DETERMINATION OF MYCOTOXINS IN FOODS

FOOD SCIENCE AND TECHNOLOGY SERIES

Food Science and Technology: New Research
Lorenzo V. Greco and Marco N. Bruno (Editors)
2008. ISBN: 978-1-60456-715-1

Food Science and Technology: New Research
Lorenzo V. Greco and Marco N. Bruno (Editors)
2008. ISBN: 978-1-61668-106-7 (Online Book)

The Price of Food
Meredith N. Fisher (Editor)
2009. ISBN: 978-1-60692-440-2

Food Processing and Engineering Topics
Maria Elena Sosa-Morales and Jorge F. Velez-Ruiz (Editors)
2009. ISBN: 978-1-60741-788-0

Food Processing and Engineering Topics
Maria Elena Sosa-Morales and Jorge F. Velez-Ruiz (Editors)
2009. ISBN: 978-1-91998-422-8 (Online Book)

Traditional Chinese Foods: Production and Research Progress
Li Zaigui and Tan Hongzhuo
2009. ISBN 978-1-60692-902-5

Traditional Chinese Foods: Production and Research Progress
Li Zaigui and Tan Hongzhuo
2009. ISBN: 978-1-61668-277-4 (Online Book)

Food Science Research and Technology
Isaak Hülsen and Egon Ohnesorge (Editors)
2009. ISBN 978-1-60741-848-1

Food Safety, Fresh Produce and FDA Oversight
Russell P. Barton (Editor)
2010. ISBN: 978-1-60692-788-5

Industrial Fermentation: Food Processes, Nutrient Sources and Production Strategies
Jürgen Krause and Oswald Fleischer (Editors)
2010. ISBN: 978-1-60876-550-8

Food Hydrocolloids: Characteristics, Properties and Structures
Clarence S. Hollingworth (Editor)
2010 ISBN: 978-1-60876-222-4

Food, Diet and Health: Past, Present and Future Tendencies
Raquel Pinho Ferreira Guine (Editor)
2010. ISBN: 978-1-60876-012-1

Sweet Potato: Post Harvest Aspects in Food, Feed and Industry
Ramesh C. Ray and K.I. Tomlins (Editors)
2010. ISBN: 978-1-60876-343-6

Microbial Interaction in Fermented Beverages
Marta E. Farías and Oscar A. Sosa; Lucía M. Mendoza, and Pedro A. Aredes Fernández
2010. ISBN: 978-1-60876-785-4

The Olive: Its Processing and Waste Management
José S. Torrecilla
2010. ISBN: 978-1-60876-719-9

Liquid Chromatography for the Determination of Mycotoxins in Foods
R. Romero-González, J. L. Martínez Vidal, and A. Garrido Frenich
2010. ISBN: 978-1-60876-882-0

Applications of Natural Products in Food
Supayang Piyawan Voravuthikunchai and Wumi Ifesan
2010. ISBN: 978-1-60876-998-8

Applications of Natural Products in Food
Supayang Piyawan Voravuthikunchai and Wumi Ifesan
2010. ISBN: 978-1-61668-370-2 (Online Book)

Extraction of Metabolities from Plant Tissues
A. Mohdaly, and I. Smetanska
2010. ISBN: 978-1-61668-252-1

Extraction of Metabolities from Plant Tissues
A. Mohdaly, and I. Smetanska
2010. ISBN: 978-1-61668-616-1 (Online Book)

Food Hydrocolloid Edible Films and Coatings
O. Skurtys, C. Acevedo, F. Pedreschi, J. Enronoe, F. Osorio, and J.M. Aguilera
2010. ISBN: 978-1-61668-269-9

Food Hydrocolloid Edible Films and Coatings
O. Skurtys, C. Acevedo, F. Pedreschi, J. Enronoe, F. Osorio, and J.M. Aguilera
2010. ISBN: 978-1-61668-740-3 (Online Book)

Fermented Milk Products
R. Ahmed Abdelrahman, A.M Adel and I. Smetanska
2010. ISBN: 978-1-61668-299-6

Fermented Milk Products
R. Ahmed Abdelrahman, A.M Adel and I. Smetanska
2010. ISBN: 978-1-61668-741-0 (Online Book)

FOOD SCIENCE AND TECHNOLOGY SERIES

LIQUID CHROMATOGRAPHY FOR THE DETERMINATION OF MYCOTOXINS IN FOODS

R. ROMERO-GONZÁLEZ
J. L. MARTÍNEZ VIDAL
AND
A. GARRIDO FRENICH

Nova Science Publishers, Inc.
New York

Copyright © 2010 by Nova Science Publishers, Inc.

All rights reserved. No part of this book may be reproduced, stored in a retrieval system or transmitted in any form or by any means: electronic, electrostatic, magnetic, tape, mechanical photocopying, recording or otherwise without the written permission of the Publisher.

For permission to use material from this book please contact us:
Telephone 631-231-7269; Fax 631-231-8175
Web Site: http://www.novapublishers.com

NOTICE TO THE READER

The Publisher has taken reasonable care in the preparation of this book, but makes no expressed or implied warranty of any kind and assumes no responsibility for any errors or omissions. No liability is assumed for incidental or consequential damages in connection with or arising out of information contained in this book. The Publisher shall not be liable for any special, consequential, or exemplary damages resulting, in whole or in part, from the readers' use of, or reliance upon, this material.

Independent verification should be sought for any data, advice or recommendations contained in this book. In addition, no responsibility is assumed by the publisher for any injury and/or damage to persons or property arising from any methods, products, instructions, ideas or otherwise contained in this publication.

This publication is designed to provide accurate and authoritative information with regard to the subject matter covered herein. It is sold with the clear understanding that the Publisher is not engaged in rendering legal or any other professional services. If legal or any other expert assistance is required, the services of a competent person should be sought. FROM A DECLARATION OF PARTICIPANTS JOINTLY ADOPTED BY A COMMITTEE OF THE AMERICAN BAR ASSOCIATION AND A COMMITTEE OF PUBLISHERS.

LIBRARY OF CONGRESS CATALOGING-IN-PUBLICATION DATA

Available upon request.

ISBN: 978-1-60876-882-0

Published by Nova Science Publishers, Inc. † *New York*

CONTENTS

Preface		ix
Chapter 1	Introduction	1
Chapter 2	Presence of Mycotoxins in Food	3
Chapter 3	Sample Preparation	9
Chapter 4	Chromatographic Methods for Mycotoxin Analysis	21
Chapter 5	Conclusion	55
References		57

PREFACE

Despite of the intentional use of chemicals, nowadays food contamination due to natural toxicants is also a mayor public concern. Mycotoxins are secondary metabolites produced by many species of fungi and they are one of the major contaminants of agricultural products. Because their high toxicity, many countries have set up regulations for their control in foods of plant origin for human. In order to fulfil the requirements described in these regulations, sensitive, selective and reliable analytical methods have been developed in order to determine as many mycotoxins as possible in one single analysis.

Traditionally mycotoxins are mainly determined by immunoassay screening methods or by single compound chromatographic analytical methods, based on immunoaffinity column cleanup followed by a separation step using thin layer chromatography (TLC), gas chromatography (GC) or liquid chromatography (LC), which were coupled to conventional detectors such as electron capture detection (ECD), fluorescence or UV-visible detection. In some cases, especially when fluorescence detection was used, it was necessary to include a pre or post-column derivatization step in order to increase the detection capabilities of the analytical method. However, the application of hyphenated chromatographic techniques, especially LC coupled to mass spectrometry (MS) and LC-MS/MS, has several advantages including simple treatment, due to further clean up procedures with immunoaffinity columns can be avoided, rapid determination and high sensitivity. Furthermore, they can be used for the simultaneous determination of several types of mycotoxins in contaminated food samples, although the number of developed methods is still limited.

Despite of the determination of regulated mycotoxins, it is important to detect mycotoxins for which standards are not commercially available as well as metabolites produced by fungi involved in food spoilage, and LC coupled to other type of analyzers such as time of flight (TOF), provides valuable information related to these compounds.

An overview of liquid chromatographic methods, providing general aspects regarding the determination of mycotoxins in food and emphasizing the use of hyphenated techniques is presented. Advantages and disadvantages of chromatographic techniques are also evaluated, indicating the new advances which are mainly focused on the reduction of the analysis time. These advances allow the application of the developed methods for the determination of multi-component mycotoxins in routine analysis or in monitoring programs, in which a large number of samples must be analyzed.

Chapter 1

INTRODUCTION

The contamination of food by the intentional use of chemicals, such as pesticides, polychlorinated biphenyls or veterinary drugs, is a worldwide public health concern. However, food contamination due to natural toxicants, as mycotoxins, can also compromise the safety of food and feed supplies and adversely affect health in humans and animals [1].

Mycotoxins are toxic natural secondary metabolites produced by many species of filamentous fungi, such as *Fusarium*, *Aspergillus* and *Penicillum* [2], on agricultural commodities in the fields or during postharvest (transport, processing and storage) [3]. They are more than three hundred different mycotoxin species that have been discovered, although the most predominant mycotoxins are the aflatoxins produced by *Aspergillus* species, ochratoxin A and patulin produced by both *Aspergillus* and *Penicillium* species, and toxins from fungi belonging to *Fusarium* such as deoxynivalenol, zearalenone, T-2 and HT-2 toxins, and fumonisins. They are ubiquitous compounds that can be found in several human foods like cereals [4], coffee [5], juices [6], milk [7], beer [8] and wine [9].

These compounds have been considered as the most important chronic dietary risk factor, higher than synthetic contaminants, plant toxins, food additives or pesticide residues [10]. Due to the health risks for humans and animals, international organizations have established maximum levels of mycotoxins in foodstuffs in order to assure food safety [11,12].

In order to ensure compliance with current legislation it is necessary to provide reliable and accurate mycotoxin analytical methods which allow unambiguous identification and confirmation, as well as an accurate quantification at trace levels, due to these compounds are toxic at very low concentrations.

Furthermore the chemical diversity of the mycotoxins and the wide range of agricultural commodities and foods pose a challenge to method development. In order to achieve this goal, several analytical methods have been developed. Immunological techniques based on specific monoclonal and polyclonal antibodies produced against several toxins are commercially available [13], and they have become a standard tool for rapid screening of mycotoxins. However they fall short in providing a definitive confirmation of the toxin and an accurate quantitative determination. Furthermore, one of the tendencies in current analytical chemistry is the simultaneous determination of multiple analytes in one single run for a given sample, and for that purpose, chromatographic methods such as gas chromatography (GC) [14], liquid chromatography (LC) [15] or thin layer chromatography (TLC) [16] are more suitable. Besides, they give better performances than immunoassays techniques in terms of trueness, precision and sensitivity. In fact, most of the validated official methods for mycotoxin detection are based on chromatographic principles [17]. Due to mycotoxins are not very volatile, they must be derivatized if GC is used, whereas this step can be avoided when LC is applied. Depending on the nature of the mycotoxin, several detectors such as electron capture detection (ECD), UV or fluorescence can be coupled to chromatographic techniques for the determination of these compounds [14,15]. Nevertheless, when LC coupled to fluorescence detection is used, a post-column derivatization is usually required in order to detect some target compounds [18]. These conventional detection techniques have been replaced by mass spectrometry (MS) detection [19], bearing in mind that public health agencies rely on detection by MS for unambiguous confirmation [20]. MS offers high sensitivity and selectivity, no derivatization is required and identification and confirmation can be carried out in one single step, minimizing sample extraction and clean-up procedure.

Due to the complexity of the samples, a pre-treatment step is mandatory before chromatographic analysis in order to minimize the effect during chromatographic separation. Basically, solvent extraction or immunoaffinity columns can be used in order to isolate and/or preconcentrate mycotoxins, allowing the determination of these compounds at concentrations below μg/L or μg/kg [2].

An overview of the analytical methods based on chromatographic techniques for the determination of mycotoxins in food and feed is provided, showing their advantages and shortcomings, as well as the future trends in this topic, mainly based on a single extraction of mycotoxins and direct injection into the chromatographic system coupled to tandem mass spectrometry (MS/MS).

Chapter 2

PRESENCE OF MYCOTOXINS IN FOOD

Mycotoxins are toxic compounds produced by several moulds in many foodstuffs under particular environmental conditions. Their presence depends on several factors such as fungal strain, climate and geographical conditions, cultivation technique and foodstuff conservation [21]. To date, several hundred mycotoxins have been identified, deriving from approximately 200 different fungi [17]. However, only a limited number (about 20) frequently occurs at significant concentrations in food and foodstuffs. Among these compounds, basically, ochratoxin A, aflatoxins, tricothecenes and fumonisins, are mainly detected, showing in Figure 1 the structure of the most common mycotoxins detected in food samples.

Ochratoxin A is a mycotoxin with nephrotoxic and immunosuppressive properties, and the International Agency for Research on Cancer (IARC) classified it as a possible carcinogen to humans [22]. It has been found in several food commodities such as wine, coffee, cereals and beer [23]. In cold climates, ochratoxin A is mainly produced by *Penicillum verrocosum* or *Penicillum nordicum*, whereas in tropical and subtropical regions, filamentous fungi belonging to *Aspergillus* are responsible for ochratoxin A production in grapes [24], provoking its presence in wine [25]. This mycotoxin has been detected in wines ranging from 0.5 µg/L to 15.6 µg/L [26], although in southern regions of Europe and in North Africa, higher concentrations were detected in red wine [27]. In general, red wines contain higher amounts of ochratoxin A than rose and white wines, which indicates a clear relationship between the process of maceration and ochratoxin A solubilization in the grape must. Furthermore, it can be suggested that it is stable in wines for at least one year, and the concentration of ochratoxin A in wine depends on the meteorological conditions [28].

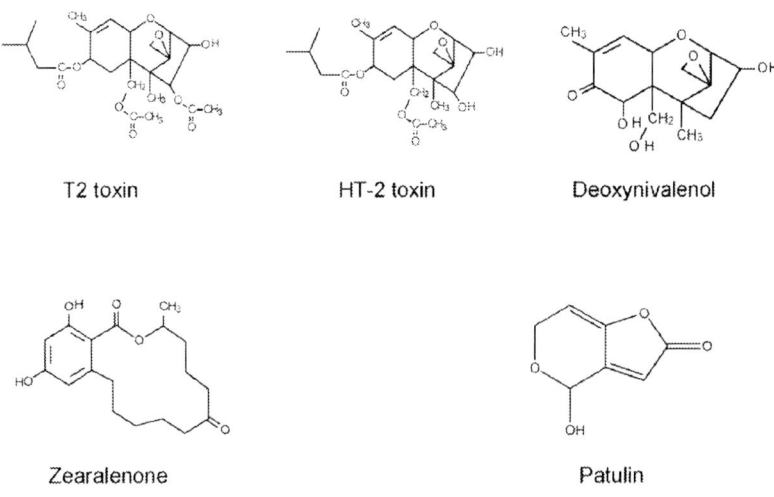

Figure 1. Chemical structure of the most common detected mycotoxins in food samples.

Codex Alimentarius Commission has suggested that 15 % of the total intake of ochratoxin A in humans is due to wine consumption [29]. In another study carried out by the European Commission, it can be noted that cereals contribute to 44 % of the total amount of ochratoxin A consumed by the EU population, whereas beverages such as wine, coffee and beer contribute to 10, 9 and 7 % respectively. Ochratoxin A can also be detected in coffee (green and processed), and several authors have reported that the contamination in green coffee ranges from 0.2 to 360 µg/kg [30]. Furthermore, mouldy cereals, mostly invaded by *Penicillum* and *Aspergillus* were often contaminated with ochratoxin A (12.8 %) and also with other mycotoxins to a lesser extent [31].

Aflatoxins are mainly produced by *Aspergillus flavus* and *Aspergillus parasiticus* and they can contaminate raw material such as ground nuts, pistachios, maize, wheat and other food commodities derived from these [32]. Aflatoxin B_1 has the highest toxicity, and epidemiological studies have indicated that prolonged exposure to aflatoxin B_1 can provoke liver cancer [33], and it has been classified as a human carcinogen [34]. Aflatoxin M_1 is the major metabolite of aflatoxin B_1 produced by ruminants and it can be detected in urine, blood, milk and internal organs of animals ingesting aflatoxin B_1-contaminanted feed [35]. For instance, cows could excrete milk with up to 0.05 µg/L of aflatoxin M_1 if their daily intake of aflatoxin B_1 reaches 70 µg [36]. Furthermore, this is the mycotoxin of major concern in milk, due to its toxicity and carcinogenic properties [37]. Because of the binding of aflatoxin M_1 to the protein fraction, in particular the association with casein, it can also be present in dairy products [38].

Another important family of mycotoxins is the trichotecenes, which are mainly produced by *Fusarium*. They are a group of over 148 mycotoxins called sesquiterpenoids, which contain an olefinic group linking carbons, an epoxide between carbons and a variable number of hydroxyl and acetoxy groups [39]. Many of the toxic properties of these mycotoxins have been attributed to the epoxy group, and they can inhibit protein, DNA, RNA synthesis and damage cellular membranes [40]. They are divided into four groups depending on their molecular structure and most attention has been paid to group A and B. Group A is the less polar of the four groups, and it includes T-2 and HT-2 toxins. Group B, which differs from group A in the possession of a keto group at position C-8, includes deoxynivalenol, nivalenol, 3-acetyldeoxynivalenol and 15-acetyldeoxynivalenol. Although group D is the most toxic of all known trichothecenes, it rarely occurs in food and feed. Deoxynivalenol is often used as an indicator of *Fusarium* infection, although other trichothecenes such as nivalenol, T-2 and HT-2 toxins are usually present in food as well [41]. These mycotoxins have also been detected in wheat-products such as bread and pasta, which are the predominant source of deoxynivalenol, nivalenol, T-2 and HT-2 toxin [42].

Other species produced by *Fusarium* are zearalenones and fumonisins [43], and their prevalence mainly depends on weather conditions during plant's growing period.

Fumonisins are mainly produced by *Fusarium monoliforme* and *Fusarium proliferatum*, and they are characterized by two tricarballylic acid side chains esterified to a 20-carbon backbone possessing a primary amine and hydroxyl functional groups. Among the 15 fumonisin analogues isolated, only three, fumonisins B_1, B_2 and B_3 appear to be of major importance, given the occurrence as natural contaminants of more than half of the corn and corn-based products worldwide [44], being fumonisin B_1 typically found at the highest levels [43]. Fumonisins can cause leukoencephalomalacia and pulmonary edema in swine [45], and the consumption of fumonisin contaminated maize has been associated with the incidence of cancer in rural areas [46]. Despite of maize, it can be detected in rice, wheat, millet and sorghum [32].

Zearalenone is a derivative of α-resorcylic acid lactone and it mainly occurs in infected corn or wheat together with trichothecene mycotoxins. It can cause disturbances and alterations in genital organs [41].

An important mycotoxin detected in apple and apple juices is patulin, which is produced by *Penicillum* species. It has been shown to be immunosuppressive, it is capable of disturbing the mitochondrial and plasma membrane functions, as well as, it inhibits the activity of numerous enzymes [47]. Furthermore, chronic exposure to patulin has been shown to induce the formation of cancerous tumours and genetic mutations [48].

Although specific mycotoxins can be detected in some foods, such as ochratoxin A in wines, aflatoxin M_1 in milk and patulin in apple juice, cereals are particularly subject to fungal infection and subsequent multi-mycotoxin contamination. This is greatly affected by environmental factors such as moisture content of substrate, temperature, grain species and soil type [49]. Cereals are mainly infested by *Fusarium*, and deoxynivalenol was the predominant toxin, and its concentration can range from 5 to 900 µg/kg [39], although other type of mycotoxins such as trichothecenes, fumonisin B_1, aflatoxins and ochratoxin A can be detected in cereals and final products such as breakfast cereals [32].

2.1. LEGISLATION

Regulations have been set by national and international organizations such as the European Union, the US Food Drug Administration (FDA), the World Health Organization (WHO) and the Food and Agriculture Organization (FAO) of the United Nations, in order to restrict the intake of mycotoxins [50]. However the legal limits vary significantly from country to country and they depend on mycotoxin type and matrix [17]. European Union has established regulatory limits for controlling aflatoxins B_1, B_2, G_1 and G_2 in cereals, nuts, nut products and dried fruit, aflatoxin M_1 in milk and ochratoxin A in cereals [11]. For instance, maximum content of 0.05 µg/L of aflatoxin M_1 has been established in milk, whereas for baby food or special dietary foods for medical purposes, the limit was only 0.025 µg/L. For the rest of aflatoxins, specific maximum levels have been provided for aflatoxin B_1, ranging from 0.1 to 8.0 µg/kg and for the sum of aflatoxin B_1, B_2, G_1 and G_2. In relation to trichothecenes, no maximum levels have been established, except for deoxynivalenol, establishing a maximum level of 200 µg/kg in baby food, whereas they established higher limits for several types of cereals and corns. European Union has set a maximum level of 400 µg/kg for the sum of fumonisin B_1 and B_2 in maize-based foods. In processed cereal based foods and baby foods, more restrictive limits of 200 µg/kg have been set [51].

European legislation has established maximum levels of ochratoxin A in several matrices, such as cereals, coffee and wine, and it recommends an ochratoxin A tolerance level lower than 2.0 µg/kg for all types of wine. On the other hand, WHO has set a provisional tolerable weekly intake level for ochratoxin A at 100 ng/kg of body weight [52].

In relation to patulin, the Joint Food and Agriculture Organization/World Health Organization Expert Committee on Food Aditives (JECFA) has established a provisional maximum tolerable daily intake (PMTDI) for patulin of 0.4 µg/kg of body weight/day [53]. Although WHO set a maximum permissible concentration (MPC) of 50 µg/kg, in European Countries, the maximum level range from 10 to 50 µg/kg, depending on the matrix.

In summary, it is a priority of national and international organizations to establish maximum allowed limits in food and feed and a duty of the producers to comply with them. Thus, the availability of reliable analytical methods for monitoring the presence of mycotoxins along the food chain is of the utmost importance for keeping mycotoxin contamination under control.

Chapter 3

SAMPLE PREPARATION

Complex general operations, which include sampling, sample preparation, extraction, purification and concentration of the extract, must be carried out before the determination of mycotoxins in food samples. For instance, the selection of a representative sample from the population for analysis (sampling) is one of the most important aspect of analysis, since the possible heterogeneity of the material and the potential presence of contaminants in the delivery systems [54].

For the determination of mycotoxins in foodstuffs by chromatographic techniques, extensive or selective sample preparation is required to remove the major part of matrix compounds that may interfere during the detection step. Furthermore a preconcentration step is also needed to reach the required detection limits. For that purpose immunoaffinity clean-up is usually carried out, although other alternatives can be applied.

Mycotoxins are compounds with different physicochemical properties, and normally, specific extraction and clean-up procedures have been optimized for one target mycotoxin, although in the last few years, generic extraction methods have been developed for the simultaneous extraction of several mycotoxins or group of mycotoxins [55]. However this is a difficult task, because although most mycotoxins are polar compounds, there are differences in their polarity. Ochratoxin A is less polar compound, and nivalenol and deoxynivalenol, with four and three hydroxyl groups respectively, are more polar. In this chapter, the most common extraction procedures in liquid and solid samples will be discussed.

3.1. Extraction of Mycotoxins from Liquid Samples

In general the extraction of mycotoxins from liquid samples such as wine, beer, juices and milk, is based on solid phase extraction (SPE) using immunoafinity columns or other type of sorbents. However other extraction techniques such as liquid-liquid extraction (LLE) [56], solid phase microextraction (SPME) [57] and liquid phase microextraction (LPME) [58] have been applied as it can be observed in Table 1.

In wine, ochratoxin A has been extracted using several organic solvents such as toluene or chloroform [59], although the most common procedure is based on the extraction with hydrogen carbonate and polyethylene glycol (PEG) solution [60,61]. As a clean-up step, the application of immunoaffinity columns has been widely used because it allows the isolation of the analyte from the matrix interferences, retaining compounds that create interference in most of the analytical methods [62]. Because its specificity and analyte preconcentration, detection limits lower than 0.01 µg/kg can be obtained [63]. In fact, for the determination of ochratoxin A in wines, an official method based on dilution with hydrogen carbonate and PEG solution followed by immunoaffinity columns clean-up, has been proposed [64], observing that ethanol and glucose content did not interfere with the clean up of ochratoxin A by immunoaffinity column. These immunoaffinity columns have become very popular in mycotoxin analysis as a very selective and time-saving one-step sample clean-up tool that facilitates high extraction capacity and an almost complete removal of matrix compounds. Furthermore, automate methods have been developed, and a robotic sample processor has been used [65] in order to minimize sample handling and increase sample throughput.

A comparison of different extraction methods is shown in Figure 2, where it can be observed that the best extraction method is based on immunoaffinity column clean-up, whereas the dilution with Na_2CO_3 and NaCl or $NaHCO_3$ and NaCl provides worse results. LLE can provide suitable results (Figure 2c), but it fails when ochratoxin A must be extracted from red wines [60].

Although these columns have been used as a clean-up and enrichment steps by several methods validated by European Union and AOAC International [32], they present some problems such as the complex matrices contain thousands of compounds, and some of them may be able to interfere with the antibodies, limiting the column capacity for the adsorption of the toxin and the composition of the matrices may interfere with the toxin structure making them not extractable and/or not recognisable by the antibodies.

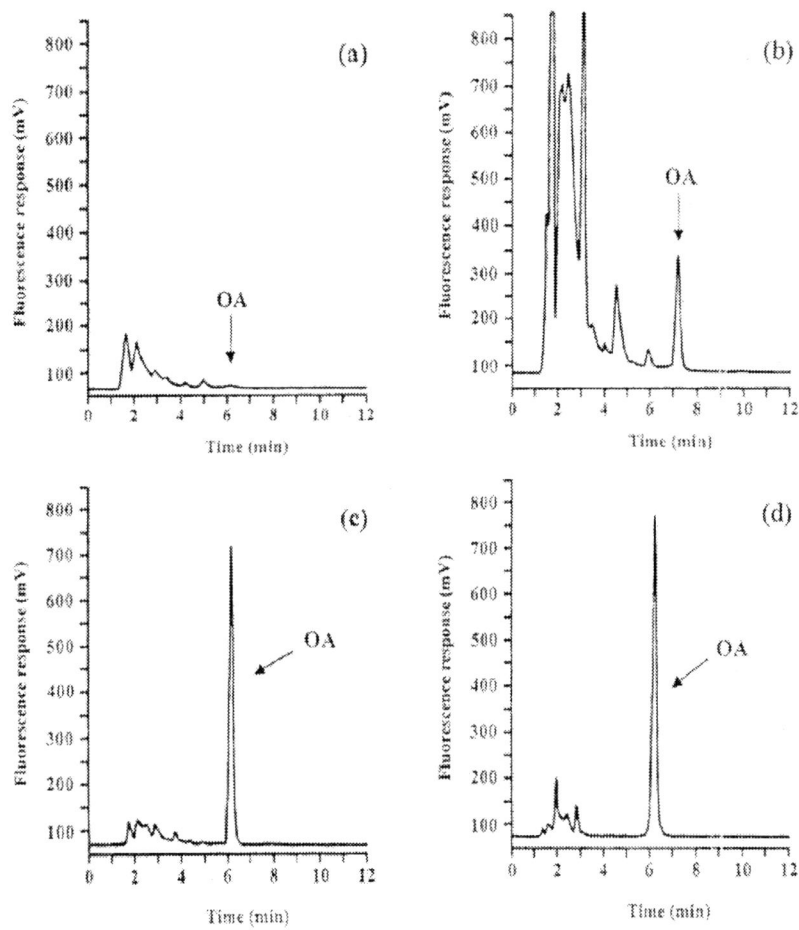

Figure 2. Chromatograms relevant to a red wine sample (injected amount equivalent to 2 mL wine) naturally contaminated with 4.72 ng/mL ochratoxin A (OA) using different sample preparation (see Reference [60] for details) prior to immunoaffinity clean-up: (a) dilution with Na_2CO_3+NaCl solution; (b) dilution with $NaHCO_3$+NaCl solution; (c) liquid–liquid extraction and (d) dilution with 1% PEG 8000+5% $NaHCO_3$ solution, filtration and immunoaffinity column clean-up. Fluorescence detection (excitation wavelength: 330 nm; emission wavelength: 460 nm). Reprinted from [60], copyright 1999, with permission from Elsevier.

In order to reduce the cost of the analysis, alternatives to immunoafinity columns have been checked, using typical sorbents during SPE, as OASIS HLB, C18, and ion exchangers (see Table 1).

Although immunoaffinity columns show better performance characteristics, anion exchange and C18 columns are less expensive and easier to handle, being a good alternative to immunoaffinity columns [66]. Furthermore, this extraction step can also be automated, and in general, on-line SPE provides better precision, sensitivity and higher sample throughput than off-line SPE [67]. When anion exchange is compared with C18, it can be noted that whereas C18 is more an enrichment than clean-up of wine samples, anion exchanger can remove more interferences [68].

Patulin can be extracted from apple juice applying conventional LLE, using ethylacetate as organic solvent [69,70]. In order to analyze lower concentration of patulin, the organic phase is usually evaporated and the residue is dissolved in water at pH 4 prior analysis. However, other alternatives based on SPE have been applied, using several sorbents. For instance, conventional C18 or OASIS HLB have been applied [71,72], although alternative sorbents such as polyvinylpolypyrrolidone-octadecyl [72] or styrene-divinylbenzene with N-methacrylate [48] can be used. In this case, the extraction procedure only takes 10 min to extract, isolate and purify this mycotoxin.

For the extraction of aflatoxin M_1 in milk, acetonitrile and *n*-hexane can be used. Acetonitrile was used to precipitate milk proteins and *n*-hexane removes fat [73], followed by a clean-up step using OASIS HLB cartridges. However, it can be noted that if fat globules were removed by centrifugation, low recovery of other mycotoxins such as zearalenone, was obtained [37]. In some cases, it was observed that milk was only centrifuged prior the application of immunoaffinity column [74,75]. Furthermore, it can be indicated that despite of aflatoxin M_1 other mycotoxins, such as fumonisins, trichothecenes, and ochratoxin A, can be extracted from milk [37].

Finally it can be observed that aflatoxins or trichothecenes can also be extracted from olive oil [76] and maize oil [40]. For this type of matrix, SPE can not be used due to the difficulties to pass the oil through the SPE cartridge, and other extraction techniques such as conventional LLE or matrix solid phase extraction must be used.

Table 1. Extraction procedures for the determination of mycotoxins in liquid samples[1]

Mycotoxin	Matrix	Sample treatment	Observations	Ref
Ochratoxin A	Wine	SPME: PDMS/DVB	Extraction time: 60 min	[57]
	Wine	LPME. Organic and acceptor phase: 1-octanol	Extraction time: 120 min	[58]
	Wine	5 mL of wine + 5 mL of CHCl$_3$	Clean-up with IAC (Easy-extract)	[59]
	Wine	10 mL of sample + 10 mL of solution containing NaHCO$_3$ (5 %) and PEG (1 %)	Clean-up with IAC (OchraTest)	[60]
	Wine	IAC (Ochraprep). Elution with methanol	Dilution (1:1) with phosphate buffered saline prior IAC	[62]
	Wine	5 mL of wine diluted with 60 mL of phosphate buffered saline, pH 8.2	Clean up with IAC (automated system)	[65]
	Wine	Anion exchange extraction (Oasis MAX)	Elution with two portions of 0.5 mL methanol–acetic acid (98/2)	[68]
	Wine, beer	On-line SPE. C18 cartridge	Elution with methanol and water	[67]
	Wine, must, beer	SPE: C18 cartridge	Elution with methanol/acetic acid (99.5:0.5)	[66]
	Beer	10 mL of sample + 10 mL of solution containing NaHCO$_3$ (5 %) and PEG (1 %)	Clean-up with IAC (OchraTest)	[61]

Table 1. (Continued)

Mycotoxin	Matrix	Sample Treatment	Observations	Ref
	Beer	LLE: Chloroform	Addition of lead hydroxyacetate to precipitate some bulk components	[56]
Aflatoxins and Ochratoxin A	Beer	SPE: Oasis HLB	Elution with acetonitrile	[8]
Patulin	Apple juice	SPE with Aqusis PLS-3 (styrene-divinylbenzene with N-methacrylate)	Elution with methanol	[48]
	Apple juice	20 mL of sample + 20 mL of ethylacetate	Evaporation of the organic phase and redissolved in water at pH 4	[69]
	Apple juice	10 g of sample + 10 mL of ethylacetate (twice)	Evaporation of the organic phas and redissolved in water at pH 4	[70]
	Apple juice	SPE: C18 cartridges	Elution with hexane: ethyl acetate: acetone (1:5:4, 1:4:5, 1:3:6, respectively)	[71]
	Apple juice	SPE: Tandem PVPP-C18 or Oasis HLB	Elution with diethylether	[72]
Aflatoxin M_1	Milk	SPE with C18. Elution with CH_2Cl_2/acetone (95:5)	Dilution with water (1:1) prior SPE	[38]
	Milk	SPE with C18 disk	Clean up with IAC (Afla M1)	[73]
Aflatoxins	Milk	40 g of milk centrifuged at	Clean up of skimmed milk	[74]

Mycotoxin	Matrix	Sample Treatment	Observations	Ref
		3500 rpm	with IAC (RIDA aflatoxin)	
	Milk	Centrifuge milk at 2000g	Clean up of skimmed milk with IAC (AflaStarM1)	[75]
18 mycotoxins	Milk	5 mL of sample + 10 mL hexane + 16 mL of CAN	Clean up with OASIS HLB	[37]
Aflatoxins	Olive oil	MSPD: Sorbent C18. Elution with methanol/water (80:20)	No clean up was necessary	[43]
Trichothecenes A, B and D	Maize oil	LLE: Methanol	Oil extracted three times	[40]

[1]Abbreviations: DVB: Divinhylbenzne; IAC: Immunoaffinity column; LLE: Liquid-liquid extraction; LPME: Liquid phase microextraction; MAX: Mixed-mode anion-exchange and reversed-phase sorbent; MSPD: Matrix solid phase dispersion; PDMS: Polydimethylsiloxane; PEG: Polyethylene glycol; PVPP: polyvinylpolypyrrolidone-octadecyl; SPME: Solid phase microextraction.

3.2. Extraction of Mycotoxins from Solid Samples

For the analysis of solid samples such as cereals, nuts, biscuits or other type of foodstuffs, prior extraction step, a reduction of the particle size is necessary in order to increase the surface area and decrease the mean free path of extraction solvent.

Generally, the extraction of mycotoxins from solid samples is based on the solid liquid extraction with a solvent or mixture of solvents followed by a clean-up procedure. Various combinations of solvents have been used for extraction of mycotoxins from grains, foodstuffs and other solid materials (Table 2). Acetonitrile/water have been mainly used as extractant solvent by several laboratories in different ratios [77,78], although other mixtures such as methanol/water [79,80] or chloroform/water [15] have been also used. The best solvent mixture depends on the polarity of the extracted mycotoxins. Most of the mycotoxins can be extracted when a mixture of acetonitrile/water 80:20 (v/v) is used. However, for the quantitative extraction of polar mycotoxins such as nivalenol, deoxynivalenol or fumonisins, higher percentage of water are used in order to provide better recoveries [74,81-84].

Due to the amount of coextractives typically contained in solid matrices, such as lipids and pigments, a clean-up procedure must be applied. Several procedures have been developed for the clean-up and enrichment of the extract based on SPE. Immunoaffinity columns, which are very specific, have been used for the extraction of several types of mycotoxins, such as fumonisins [85], trichothecenes [42], aflatoxins [86], zearalenone [87] and ochratoxin A [26]. Some works compared different immunoaffinity columns, observing that in general, MultiSep 226 provides better results than MultiSep 227, specially for the extraction of aflatoxins and zearalenone [86], because it allows compounds to pass through over a wider range of polarities. However, the main drawback is that more matrix compounds can also be found.

However, other type of sorbents such as C18 [38], strong anion exchanger [88], graphitized carbon black, commercially named carbograph [39,89], florisil [15] or amino have been also tested [90], providing reliable results for the simultaneous determination of several classes of mycotoxins. In this sense, C18 has been proposed as an interesting alternative to *n*-hexane partitioning for the defatting of the matrices, when mycotoxins are extracted from grains [89]. In relation to carbograph, it can behave both as reversed phase and as anion-exchanger sorbent increasing the possibilities of an efficient clean up procedure [43]. In some cases, specific home-made cartridges can be developed. For instance, secondary rabbit anti-mouse antibodies can be coupled to CNBr-activated Sepharose 4B to isolate ochratoxin A from roasted coffee [79].

It is important to indicate that when immunoafinity columns are used for the determination of ochratoxin A in breakfast cereals, it was observed that ochratoxin amount decreased with increasing the extraction pH. This can be due to the extraction from a neutral medium gives lower recoveries than from an acidic medium and in alkaline medium, ochratoxin A is converted to an open ring which could not be recognised by the antibodies, and $NaHCO_3$ is usually used to work within a suitable pH range [26].

Nowadays, multi-mycotoxin methods are being developed and sample treatment is a critical step, because significant losses may occur during the extraction or clean-up. Furthermore, sample treatment should be kept as simple as possible, and clean-up can be avoided, specially when chromatographic techniques have been coupled to MS detection [56,91]. These methods are based on solvent liquid extraction, using a mixture of acetonitrile/water, and dilution of the extract prior chromatographic determination, eliminating further clean-up steps [77,83,92,93]. These methodologies allow the simultaneous extraction of a broad range of compounds, and they have several advantages such as the speed and ease of application with one method for multiple determination, which is of special interest for high-throughput routine analysis.

Table 2. Extraction procedures for the determination of mycotoxins in solid samples[1]

Mycotoxin	Matrix	Extraction procedure	Clean-up	Observations	Ref
Trichothecenes A, B and D	Maize	MSPD: C18	Not necessary	Elution with ACN/Methanol	[40]
Trichothecenes A, B and zearalenone	Maize	10 g of sample + 40 mL of ACN/water (84:16)	Mycosep 226	Extraction time: 90 min.	[77]
Trichothecenes	Maize	1 g of sample + 25 mL of ACN/water (75:25)	Clean up with Carbograph-4	Elution with Methanol	[39]
Fusarium toxins	Maize	1 g of sample + 10 mL ACN/water (75:25)	SPE: C18 or Carbograph	Elution with Methanol/CH_2Cl_2 (80:20)	[43]
39 mycotoxins	Maize, wheat	0.5 g of sample + 2 mL of ACN/water/acetic acid (79:20:1)	Not necessary	Dilution with ACN/water/acetic acid (20:79:1)	[83]
Trichothecenes	Wheat grain	25 g of sample + 100 mL ACN/methanol (84:16)	Not necessary	Extraction time: 120 min	[91]
Trichothecenes, fumonisins and zearalenone	Corn	1 g of sample + 10 mL of ACN/water (75:25). Extraction with C18	Clean up with Carbograph-4	Elution with 1 mL of methanol and 8 mL of CH_2Cl_2/methanol (80:20).	[89]
17 mycotoxins	Corn feed	10 g of sample + 40 mL of ACN/water (84:16)	MultiSep 226 Aflazon and Multifunctional cartridges	Addition of 40 mL of ACN prior clean-up step	[82]
Fumonisin B_1 and B_2	Corn meal	5 g of sample + 2x12.5 mL of ACN/methanol/water (30:30:40)	IAC: Fumoniprep	Elution with methanol	[85]
Fumonisin B_1, B_2, B_3, B_4	Corn samples	50 g of sample + 25 mL ACN/water (1:1)	Amberlite XAD-4	Extraction overnight	[84]
12 mycotoxins	Corn feed	5 g of sample + 10 mL of ACN/water (80:20)	Not necessary	Extraction time: 10 min.	[93]
Zearalenone	Corn samples	25 g of samples + 100 mL ACN	IAC (Easi Extract)	Elution with ACN	[87]
Trichothecenes, aflatoxins and zearalenone	Corn, wheat, cornflask, biscuits	10 g of sample + 40 mL of ACN/water (85:15).	IAC: MultiSep 226	Discard the first 3 mL of eluate	[86]

Table 2. (Continued)

Mycotoxin	Matrix	Extraction procedure	Clean-up	Observations	Ref
Trichothecenes A and B	Cereals and grains	20 g of sample + 100 mL of ACN/water (85:15)	IAC: Mycosep 226 and 227	Extraction time: 60 min	[42]
33 mycotoxins	Peanut, pistachio, wheat, maize, cornflakes, raisins, figs	25 g of sample + 100 mL of ACN/water (80:20)	Not necessary	Dilution with water (1:3) prior analysis	[78]
87 analytes	Moldy food	0.5 g of sample + 2 mL ACN/water/acetic acid (79:20:1)	Not necessary	Extraction time: 90 min. Dilution 1:1 with ACN/water/acetic acid (20:79:1)	[92]
Aflatoxin B_1, ochratoxin A	Spices	2.5 g of sample + 7.5 mL of methanol/3 % $NaHCO_3$ (80:20)	SPE: NH_2 as sorbent	Detection immunolayers	[94]
Ochratoxin A	Wine grapes	50 g of sample + 150 ml of solution (5 % $NaHCO_3$ and 1 % PEG).	IAC: OchraTest. Elution with methanol	IAC washed with 5mL of solution (2.5% NaCl, 0.5% $NaHCO_3$)	[26]
Ochratoxin A	Roasted coffee	Extraction with 15 mL of methanol/3 % $NaHCO_3$ (80:20)	Tandem assay column: secondary rabbit antibodies coupled to CNBr-activated Sepharose 4B	Reduction of methanol content by dilution before clean-up procedure	[79]
Deoxynivalenol	Egg	Extraction with ACN/water (84:16)	Not necessary	Extraction time: 120 min	[41]

Mycotoxin	Matrix	Extraction procedure	Clean-up	Observations	Ref
Zearalenone	Egg	Extraction with ACN/water (75:25)	IAC	Addition of phosphate buffered prior IAC	[41]
Aflatoxin M_1	Cheese	10 g of sample + 50 mL of CH_2Cl_2:acetone (1:1)	Clean up with C18	Evaporation of organic layer prior clean-up	[38]
Ochratoxin A	Kidney	7.5 g of sample + 37 mL of $CHCl_3$ + 1 mL of H_3PO_4	Clean up with SAX	Elution with ethylacetate/formic acid (99:1)	[88]
Zearalenone, ochratoxin A	Soil	0.5 g ascorbic acid and 30 mL MeOH:H_2O (90:10)	SPE (C8). Elution with MeOH	Ascorbic acid avoids oxidation of mycotoxins	[80]

[1] Abbreviations: ACN: Acetonitrile; IAC: Immunoafinity column; MSPD: Matrix-solid phase dispersion; PEG: Polyethylene glycol; SAX: Strong anion exchange; SPE: Solid phase extraction.

Chapter 4

CHROMATOGRAPHIC METHODS FOR MYCOTOXIN ANALYSIS

A wide number of analytical methods have been developed for mycotoxin analysis. They need to have low detection limits in order to fulfil the requirements indicated by national and/or international organizations, be specific to avoid interferences, be easily applied in routine laboratories, be economical for the laboratory and provide confirmatory test for the detected analyte. Although immunochemical methods based on enzyme-linked immunosorbent assay (ELISA) can be used for screening procedures due to speed, ease of operation, sensitivity and high sample throughput [94], they can provide erroneous results due to matrix complexity and cross-reaction.

That is why, conventional analytical methods include LC with UV or fluorimetric detection with either pre or post-column derivatization [18], which are extensively used, although other chromatographic methods such as TLC [14,16] and GC [95] coupled to ECD, flame ionization detection (FID) or MS [96] are also employed. Recent advances in analytical instrumentation have highlighted the potential of LC-MS methods, which allow the multitoxin determination and they are also used for confirmation purposes [97].

These methodologies can be used in routine laboratories and for legal enforcement of food safety legislation and for regulations in international agricultural trade, and some of them have been validated by interlaboratory collaborative studies and accepted by official authorities, such as the European Committee for Standarization (CEN), the Association of Official Analytical Chemists (AOAC International) and the International Organization for Standardization (ISO) [17].

4.1. THIN LAYER CHROMATOGRAPHY METHODS

Traditionally, TLC had been the most popular method used for the mycotoxins analysis, probably because of its low analysis costs and accessibility [16]. Several methods were developed to obtain the best results with both one and two dimensional analyses [14]. The use of TLC analysis is still popular for both qualitative and semi-quantitative purposes due to its high sample throughput, low operating cost and ease of identification of target compounds, although in the last three decades, LC or GC have been more extensively applied.

Several types of mycotoxins such as trichothecenes, aflatoxins, fumonisins and ochratoxin A have been determined using this technique (see Table 3). Silica gel plates have been mainly used as stationary phase although the potential of reversed phase TLC plates has been also evaluated [98]. In relation to the mobile phase, toluene/ethyl acetate/acetone [99] can be used for the separation of trichothecenes, although acetone can be replaced by formic acid [100]. On the other hand, tert-butyl methyl ether/methanol/water [101] or toluene/ethyl acetate/chloroform/formic acid [102] have been utilised for the separation of aflatoxins. On the other hand, if reversed phase was applied, methanol/water can be used as mobile phase in order to elute the target compounds [98]. If several types of mycotoxins are separated, stepwise gradient can be used. For instance, aflatoxins and ochratoxin A can be separated using a mixture of chloroform/acetone (9:1) first, and then a mixture of toluene/ethyl acetate/formic acid (6:3:1) [103]. The first mixture is useful for the separation of ochratoxin A from aflatoxins, whereas the second one allows the separation of aflatoxins without overlap.

To detect the mycotoxin spots on thin-layer plates, UV or fluorescence detection can be used, bearing in mind the native fluorescence of some mycotoxins such as ochratoxin A or aflatoxins, providing accuracy results for the determination of these compounds, replacing visual comparison of the intensities of sample spots with the use of densitometry.

As example, in Figure 3 it can be observed the separation of aflatoxin B_1, B_2, G_1 and G_2 using tert-butyl methyl ether, methanol and water as mobile phase. No interferences were observed, allowing a suitable determination of these mycotoxins using a simple procedure, applying absorbency densitometry as detection technique [101].

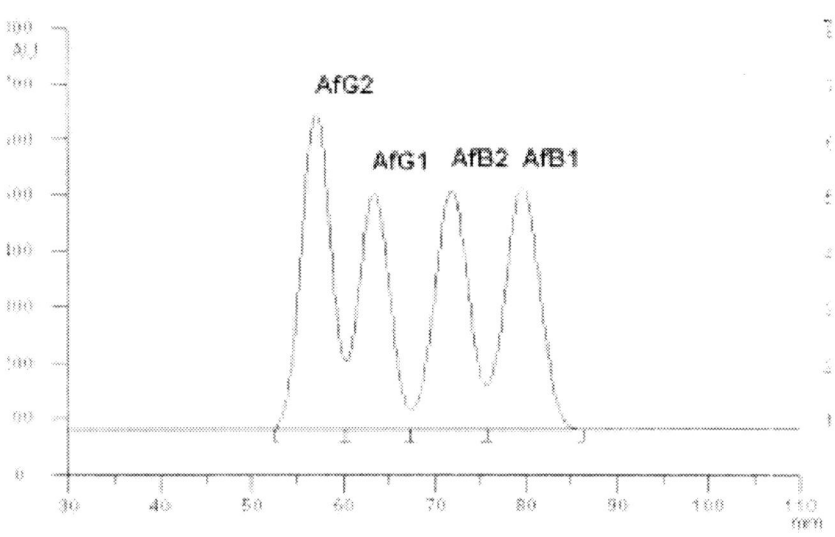

Figure 3. Chromatogram of the aflatoxins separated by TLC. Separation of aflatoxins extracted from a fortified pistachio sample (4 ng/g each aflatoxin) on a silica gel 60 TLC plate. The start position of the chromatogram was at 20 mm, while the solvent front ended at 170 mm with a run-time of 60 min. Reprinted from [101], copyright 2000, with permission from Elsevier.

Table 3. Thin layer chromatographic methods applied for the determination of common mycotoxins

Mycotoxin	Matrix	TLC procedure	Detection	Ref
Trichothecenes	Food	Silica gel plates Mobile phase: Toluene/ethyl acetate/acetone (7:2:1)	Fluorescence	[99]
Trichothecenes	Grains and animal feed	Silica gel plates Mobile phase: Toluene/ethyl acetate/formic acid (100:50:15)	Fluorescence	[100]

Table 3. (Continued)

Mycotoxin	Matrix	TLC procedure	Detection	Ref
Fumonisin B1	Maize	Reversed phase Mobile phase: methanol/aqueous 4% potassium chloride (70:30).	Visible	[98]
Aflatoxins	Corn, paprika, peanuts, pistachios	Silica gel plates Mobile phase: Tert-butyl methyl ether/methanol/water (480:15:15)	Absorbency densitometry	[101]
Aflatoxins	Corn-based food products	Silica gel plates Mobile phase: toluene:ethyl acetate:chloroform:formic acid (35:25:25:15)	UV	[102]
Aflatoxins, ochratoxin A	Cereals	Silica gel plates Mobile phase: CHCl3/acetone (9:1) and toluene/ethyl acetate/formic acid (6:3:1).	Fluorescence	[103]

4.2. GAS CHROMATOGRAPHY METHODS

GC has been applied for the identification and quantification of mycotoxins in food samples and many protocols have been developed [3] although for many of these compounds, which possess strong fluorescence or UV properties, LC methods have been more successful. In this sense, the analysis of trichothecenes by GC has been extensively studied [104] because they are not fluorescent and do not strongly adsorb in the UV-visible range, although LC can be applied for the analysis of type B [105]. Most mycotoxins are not volatile and therefore they have to be derivatized prior the injection into a GC column (see Table 4). Several techniques have been developed for the derivatisation of mycotoxins, such as silylation [106] or polyfluoroacylation [107], although in some cases this step can be avoided [108] because this derivatization step is time consuming and prone to error. In general, trichothecenes such as nivalenol and deoxynivalenol showed significantly higher abundances as trimethylsilyl derivatives whereas for trichothecenes type A, fluoracyl derivatives provide better response [91]. In this sense, Valle-Algarra et al. [109] compared different derivatization reagent, such as heptafluorobutiric anhydride (HFBA) and pentafluoropropionic anhydride (PFPA), preferring PFPA to HFBA because it provided better stability against moisture, lower cost and similar sensitivity of trichothecene derivatives, except deoxynivalenol. Figure 4 shows a GC-FID chromatogram of trichothecenes in wheat, using as derivatizating reagent trimethylsilylimidazole-bis(trimethylsilyl)acetmide-trimethylchlorosilane and FID as detector. Due to some compounds can be interfered with target analytes (Figure 4b) authors indicated that another column type must be used for confirmation in order to eliminate false positives [95].

As a general remark, these derivatization reactions need to be optimized in order to avoid multiple reaction products, as well as to remove excess of reagent [2].

In relation to ochratoxin A, LC is the technique of choice, but some papers developed GC methods, using bis(trimethylsisyl)trifluoroacetamide (BSTFA) as derivatizing reagent, which was analyzed by GC-MS in selected ion monitoring (SIM) mode by monitoring eight specific ions. However, the methodology presents poorer sensitivity, recovery and precision than LC methods [110] and it is not recommended for routine detection of ochratoxin A. In the same way, patulin can be analyzed by GC-MS [111], obtaining detection limits of 1 µg/kg. Due to the characteristics of this compound a derivatization step is necessary prior GC analysis, using N,O-bis(trimethylsilyl)trifluoroacetamide as derivatizating reagent.

Most common GC detectors such as FID or ECD (see Table 4) can be used for the determination of mycotoxins [112]. For instance, the conjugated carbonyl group in trichothecenes type B and the use of fluorine-containing derivatising agent for type A compounds make them sensitive to ECD at low detection levels [113].

However in the last few years MS has been mainly applied, using several analyzers such as single quadrupole [114], ion trap [107] and time of flight (TOF) [91]. The use of MS provides the advantage of selective, multitoxin analysis and quantitative data from a single analytical run. MS allows the detection at low concentration levels and the confirmation of the identity of the chromatographic peak by the production of characteristic fragment ions.

Recent advances in GC-MS have been developed. In this sense, comprehensive two dimensional gas chromatography coupled to time of flight (GCxGC-TOF) can be used [91]. This allows the elimination of cleanup step, taking advantage of the separation capacity of GCxGC system, simplifying the sample treatment. Furthermore, deconvolution capabilities of TOF analyzer allows the resolution of mycotoxins in complicated matrices.

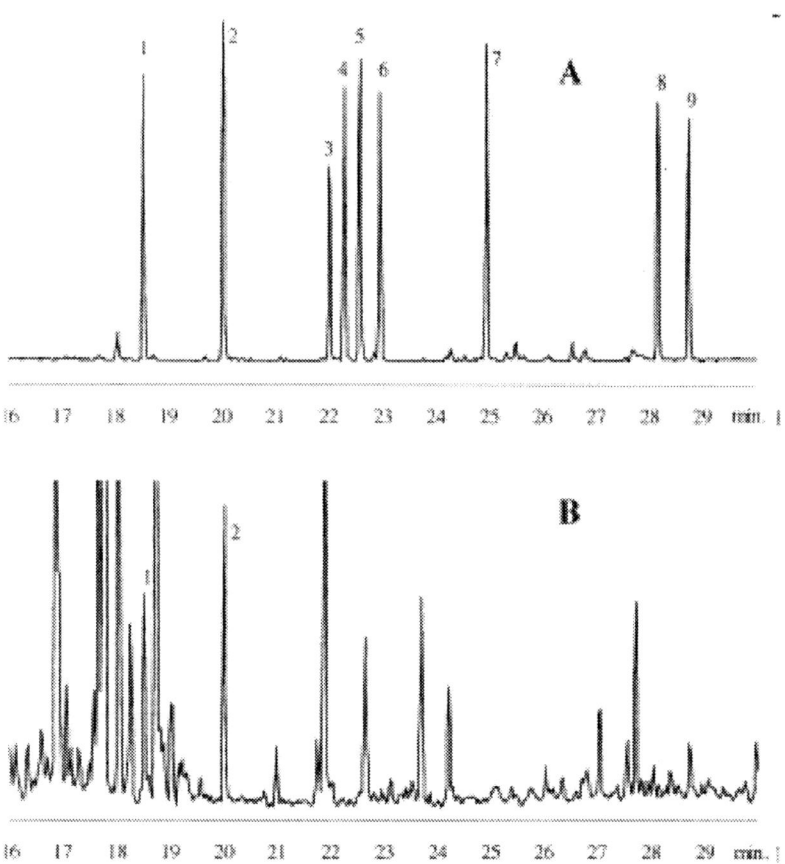

Figure 4. Chromatograms of a silylated standard solution (about 600 µg/kg) of trichothecenes (A) and of a silylated BCR CRM 379 test sample (B). Peaks: 1, α-chloralose (internal standard.): 2, deoxynivalenol: 3, 3-acetyldeoxynivalenol: 4, fusarenon X: 5, nivalenol: 6, diacetoxyscirpenol: 7, neosolaniol: 8, T-2 toxin: 9, HT-2 toxin. Type of detection: FID. Reprinted from [95], copyright 2001, with permission from Elsevier.

A final problem related to analysis of mycotoxins by GC is thermal stability, and sample gets lost when it comes into contact with the heated areas of the injector, provoking the degradation of mycotoxins and systematic errors during the determination of these compounds.

Table 4. Gas chromatographic methods for the determination of mycotoxins in foodstuffs[1]

Mycotoxin	Stationary phase	Derivatization	Detection	Detection limit	Ref
Trichothecenes B	Fused silica column HP-5 (30 m×0.32mm)	PFPA	ECD	7-28 µg/kg	[109]
Trichothecenes	CPSil 19CB (60 m x 0.25 mm)	Tri-sil-TBT	FID	25 µg/kg	[95]
Trichothecenes	HP 5 (30 m x 0.32 mm)	Tri-sil-TBT	ECD	1.6 mg/kg	[112]
Trichothecenes	NB 54 (50 m x 0.20 mm) and NB-1701 (50 mx 0.20 mm)	TFA	ECD	40-200 µg/kg	[113]
Trichothecenes	HP-5 Trace column (30 m x 0.25 mm)	Heptafluorobutyril imidazole	IT (MS/MS, NCI)	10-70 pg	[107]
Trichothecenes A, B	DB-5MS (25 m x 0.2 mm) and BPX-50 (1.3 m x 0.1 mm)	TFA and TMS	TOF	10-50 µg/kg	[91]
Trichothecenes, zearalenone	DB-5 (30 m x 0.25 mm)	TMS	Q (SIM)	5-10 µg/kg	[106]
Patulin	BPX-35 (25 m x 0.22 mm)	N,O-bis(trimethylsilyl)trifluoroacetamide	Q (SIM)	1 µg/kg	[111]
Deoxynivalenol, T-2 toxin, zearalenone	Rtx-200 (30 m x 0.25 µm)	No derivatization	Q (SIM)	0.1-0.5 mg/kg	[108]
Ochratoxin A	DB-5MS (15 m x 0.25 mm)	BSTFA	Q (SIM)	0.1 µg/L	[110]
Ochratoxin A, deoxynivalenol	OV1CB (60 m x 0.32 mm)	Deoxynivalenol derivatized with Tri-sil-TBT	Q S(IM)	5 µg/kg	[114]

[1] Abbreviations: BSTFA: Bis(trimethylsilyl)trifluoroacetamide; ECD: Electron capture detection; FID: Flame ionization detection; IT: Ion trap; NCI: Negative chemical ionization; PFPA: pentafluoropropionic anhydride; Q: Single quadrupole; SIM: Single ion monitoring; TFA: Trifluoroacetic acid; TMS: Trimethylsilyl; Tris-sil-TBT: Trimethylsilylimidazole-bis(trimethylsilyl)acetamide-trimethylchlorosilane (3:3:2); TOF: Time of flight.

4.3. LIQUID CHROMATOGRAPHY COUPLED TO CONVENTIONAL DETECTORS

Basically, most of the protocols used for LC detection of mycotoxins are very similar. The most commonly found detection methods are UV or fluorescence detectors, which rely on the presence of a chromophore in the molecules. Several mycotoxins such as ochratoxin A and aflatoxins have natural fluorescence, and can be detected directly by fluorescence detection [21,115]. However, other mycotoxins such as fumonisins lack a suitable chromophore, and their determination requires derivatization. Table 5 shows that reversed-phase column has been mainly used, and the most employed mobile phases are based on acetonitrile/water/acetic acid mixtures [56], or methanol and acidified water [80]. Sometimes, post-column addition of ammonia to acidic mobile phase has been reported to increase fluorescence yield [66]. Furthermore, isocratic mode is mainly used, but in order to reduce analysis time, gradient elution is applied (see Table 5).

As it can be noted, the method most frequently used to analyze ochratoxin A in foods are LC with fluorescence detection [26,60] although diode array detection (DAD) can be also applied [116]. DAD is usually less sensitive than fluorescence detection [66], but it can be suitable for confirmation purposes. The presence of a carboxylic group in ochratoxin A structure, with regard to chromatographic separations, implies acid or buffer addition to mobile phase to avoid peak broadening and tailing. In relation to detection conditions when fluorescence detection is used, excitation wavelength is usually fixed at 330 nm, whereas emission wavelength is usually set at 460 nm, although slight variations can be found (see Table 5).

Since aflatoxins have fluorescent properties, fluorescence detection is more sensitive than UV absorption. However, one of the main problems is that the most important aflatoxins (B_1, B_2, G_1 and G_2) fluorescence depends on the composition of the mobile phase, specially if normal phase LC was applied, reducing the sensitivity for aflatoxin B_1 and B_2 due to quenching process [18]. However if reversed phase LC is used, aflatoxins can be eluted with methanol/water or acetonitrile/water mobile phases, but the fluorescence of aflatoxin B_1 and G_1 is reduced considerably and pre or post-column derivatization is necessary in order to enhance detection. The most frequently compound for pre-column derivatization is trifluoroacetic acid forming the corresponding hemiacetals [117,118], while iodine was reported as a post-column derivatizing agent [119], which was adopted by the AOAC-IUPAC as official method for the analysis of aflatoxins [18]. However this approach present several drawbacks such as extended time to stabilize the mobile phase and iodine solution must be prepared daily for stability reasons. Bearing in mind these shortcomings, bromine has been proposed as alternative to iodine, increasing the sensitivity and recovery of aflatoxins detection [120,121]. In this sense, an electrochemical cell (Kobra cell) can be used for this derivatizion step [122], which is fitted on-line between the chromatographic system and the fluorescence detector, generating bromine from potassium bromide and nitric acid. This automated system is preferable in order to reduce the manipulation required on each sample.

In relation to other aflatoxins such as M_1, a new LC method with fluorescence detection using pyridinium hydrobromide perbromide as a post-column derivatising agent has been applied [38], achieving low detection limits in milk and cheese.

For the determination of fumonisins, fluorescence detection can be used including a derivatization step with o-phtalaldehyde [74] and naphthalene-2,3-dicarboxaldehyde [123]. However, other detection techniques such as evaporative light scattering detector (ELSD) has been successfully used to quantify underivatized fumonisin in LC analysis [84], although the obtained detection limits were higher than those obtained with fluorescence detection.

For the determination of trichothecenes, LC has been used with either DAD [124] or fluorescence detection, using pre-column derivatization with coumarin-3-carbonyl chloride, which has been mainly used as derivatizating reagent for this type of mycotoxins [125,126]. However, when this reagent is used to obtain fluorescent derivatives, an incomplete derivatisation of trichothecenes type B can occur, whereas trichothecenes type A could be derivatised more completely, which can be explained by the different polarities of the two classes of trichothecenes [127]. For the determination of these type of mycotoxins, the major disadvantage of LC is the extensive and/or very selective sample clean-up procedures required to remove matrix components that might interfere with the analyte signal and the need to preconcentrate the analyte to reach the required detection limits. However, the use of LC coupled with MS/MS can reduce the need for sample preparation, since coeluted compounds may be eliminated by MS selectivity.

Fluorescence detection is also used for the determination of zearalenone [124]. In this sense several excitation and emission wavelengths can be used, obtaining higher sensitivity when excitation wavelength of 274 nm and emission wavelength of 440 nm are used [127]. Furthermore, zearalenone can be detected simultaneously with other type of mycotoxins such as ochratoxin A, reporting detection limits of 1.0 µg/kg [80].

In relation to patulin, LC coupled with UV detection is particularly suitable, because the toxin is relatively polar and exhibits a strong absorption spectrum, although DAD can be used to distinguish patulin from co-extracted compounds [128]. Some of the original LC work involved normal phase chromatography [129], but recent publications are based on reversed phase columns [130]. Because patulin is a low-molecular mass, it is only retained on reversed-phase by the use of mobile phases with high aqueous content [131]. Basically mixtures of water/acetonitrile or water/methanol are mainly used either in isocratic or gradient mode as it can be observed in Table 5. Sometimes gradient mode must be used in order to separate patulin from interfering substances in apple juice, especially 5-hydroxymethylfurfural, which naturally occurs in apples [132], to obtain baseline separation of these two compounds taking into account that they have maximum absorbance at the same wavelength (270 nm).

In some cases it is possible the simultaneous determination of several mycotoxins such as ochratoxin A and aflatoxins, using fluorescence detectors with variable detection conditions [133], because they are different for the selected mycotoxins.

4.4. LIQUID CHROMATOGRAPHY COUPLED TO MASS SPECTROMETRY

As we indicated previously, when conventional detection is applied, specific conditions can be used for the quantification of the several classes of mycotoxins. However, the application of MS has made possible the development of multitoxin methods suitable for a range of structurally diverse toxins in one chromatographic run [81]. The need for these methods lies in the fact that a single fungal species can produce different toxins or a single agricultural commodity can be contaminated with different fungal species resulting in the co-occurrence of a number of different toxins [92]. Although the limiting factors in the use of MS as analytical tool, are the high cost of equipment, complex laboratory requirements and limitations in the type of the solvents used in extraction and separation, in the recent years this technique has became very popular, because it allows a reduction in sample treatment, it is universal, selective and sensitive detection mode. Furthermore, it increases the reliability of the obtained results. In this sense, whereas conventional detectors can produce false positive or false negative results because of retention time shifts or coelution of interfering compounds, LC-MS can be overcome this problem, combining the selectivity and structural information provided by MS with the advantages of LC. Besides, limitations such

Table 5. Liquid chromatographic methods for the determination of mycotoxins in foodstuffs using conventional detection1

Mycotoxin	Stationary phase	Mobile phase	Flow (mL/min)/Mode/ Column temperature (°C)	Detection conditions	Detection limit	Ref
Ochratoxin A	Supelco C_{18} (150 x 4.6 mm)	ACN/ H_2O/acetic acid (99:99:2)	1.0/Isocratic/RT	FL (λ_{ex}: 330 nm; λ_{em}: 460 nm)	0.01 µg/L	[21]
	Spherisorb ODS2 (250 x 4.6 mm)	ACN/ H_2O/acetic acid (99:99:2)	1.0/Isocratic/30	FL (λ_{ex}: 330 nm; λ_{em}: 460 nm)	0.004 µg/kg	[26]
	LiChrospher 100 C_{18} (250 x 4 mm)	ACN/H_2O at pH 3 (40:60)	1.4/Isocratic/30	FL (λ_{ex}: 330 nm; λ_{em}: 460 nm)	0.005 µg/L	[56]
	Supelcosil LC-18 DB (150 x 4.6 mm)	H_2O/ACN/Acetic acid (111:87:2)	1.0/Isocratic/22	FL (λ_{ex}: 332 nm; λ_{em}: 460 nm)	0.07 µg/L	[57]
	Tracer Extrasil ODS-2 (250 x 4 mm)	ACN/methanol/sodium acetate 5 mM at pH 2.2 (29:29:42)	1.5/Gradient/40	FL (λ_{ex}: 225 nm; λ_{em}: 461 nm)	0.2 µg/L	[58]
	Discovery C_{18} (150 x 4.6 mm)	ACN/ H_2O/acetic acid (99:99:2)	1.0/Gradient/NI	FL (λ_{ex}: 330 nm; λ_{em}: 460 nm)	0.01 µg/L	[60]

Table 5. (Continued)

	Column	Mobile phase	Flow rate/Elution mode/T (°C)	Detection	LOD	Ref.
	Kromasil SC-18 (150 x 4.6 mm)	ACN/ H$_2$O/acetic acid (50:49:1)	1.0/Isocratic/NI	FL (λ_{ex}: 333 nm; λ_{em}: 470 nm)	0.01 µg/L	[62]
	LiChrospher 100 C$_{18}$ (250 x 4 mm)	ACN/ H$_2$O/acetic acid (99:99:2)	1.0/Isocratic/30	FL (λ_{ex}: 330 nm; λ_{em}: 460 nm)	0.1 µg/L	[66]
	LC-NH$_2$ Supelcosil (250 x 4.6 mm)	ACN/methanol/ ammonium acetate buffer (50 mM, pH 7, (78:2:20, v/v/v).	1.0/Isocratic/RT	DAD (220-380 nm)	10 µg/L	[116]
Aflatoxin B$_1$	Phenomenex ODS (250 x 4.6 mm)	H$_2$O/methanol/ACN (6:2:2)	1.0/Isocratic/30	FL (post-column derivatization with bromine, λ_{ex}: 362 nm; λ_{em}: 425 nm)	2 ng/kg	[120]
	Uptisphere C$_{18}$ (150 x 4.6 mm)	Eluent A: 0.1 H$_3$PO$_4$ Eluent B: Methanol/ACN (50:50)	1.0/Isocratic (56:44)/25	FL (post-column derivatization with bromine, λ_{ex}: 364 nm; λ_{em}: 440 nm)	0.01 µg/L	[122]
Aflatoxin M$_1$	Supelcosil LC-18 (250 x 4.6 mm)	Acetic acid/ACN/2-propanol/H$_2$O (2:10:10:78)	1.2/Isocratic/40	FL(post-column derivatization with PBPB, λ_{ex}: 353 nm; λ_{em}: 423 nm)	1 ng/kg (milk) 5 ng/kg (cheese)	[38]
Aflatoxins	Thermo LC-Si (250 x 4.6 mm)	Toluene/ethyl acetate/formic acid/methanol	0.2/Gradient/40	FL (λ_{ex}: 365 nm; λ_{em}: 425 nm)	0.05 µg/kg	[115]

Analyte	Column	Mobile phase	Flow/Elution/Temp	Detection	LOD	Ref.
	Inertsil ODS-3 (250 x 4.6 mm)	ACN/methanol/H$_2$O (8:27:65)	0.7/Isocratic/40	FL (pre-colum derivatizacion with TFA; λ_{ex}: 365 nm; λ_{em}: 450 nm)	0.5 µg/kg	[117]
	ODS (250 x 4.6 mm)	Methanol/ACN/0.1 % H$_3$PO$_4$ (24:24:52)	1.0/Isocratic/30	FL (pre-column derivatizaciont with TFA, λ_{ex}: 364 nm; λ_{em}: 440 nm)	0.05-2 µg/L	[118]
	LiChrospher 100 RP-18e (250 x 4.6 mm)	ACN/Methanol/H$_2$O (1:1:4)	1.0/Isocratic/20	FL (post-column derivatization with iodine, λ_{ex}:360 λ_{em}: 450 nm)	0.22-0.75 µg/kg	[119]
Fumonisin B$_1$, B$_2$	Zorbax Eclipse XDB C$_{18}$ (150 x 4.6 mm)	Eluent A: Methanol Eluent B: 0.1 M phosphate buffer (pH 3.15)	0.8/Gradient/40	FL (post-column derivatization with OPA/Thiofluor, λ_{ex}: 343 nm; λ_{em}: 445 nm)	5-6 µg/kg	[85]
	X-Bridge TM C$_{18}$ (100 x 2.1 mm)	Eluent A: H$_2$O (0.5 % formic acid) Eluent B: Methanol (0.5 % formic acid)	0.3/Gradient/30	FL (pre-column derivatization with NDA, λ_{ex}: 420 nm; λ_{em}: 500 nm)	15-20 µg/kg	[123]
Fumonisin B$_1$, B$_2$, B$_3$, B$_4$	Alltima C18LL (250 x 4.6 mm)	Eluent A: H$_2$O/TFA (100:0.025) Eluent B: ACN/TFA (100:0.025)	1.0/Gradient/NI	ELSD (Drift tube temperature: 45 °C)	0.1 mg/kg	[84]

Table 5. (Continued)

Ochratoxin A, Aflatoxin B_1	Spherisorb C_{18} (250 x 4.6 mm)	0.33 M H_3PO_4/ACN/2-propanol (650:400:50)	0.5/Isocratic/NI	FL (Ochratoxin A: λ_{ex}: 335 nm; λ_{em}: 465 nm; Aflatoxin B1: λ_{ex}: 365 nm; λ_{em}: 440 nm)	Ochratoxin A: 0.22 µg/kg; Aflatoxin B_1: 0.07 µg/kg	[133]
Nivalenol, Deoxynivalenol	Eurospher 100 RP-18 (250 x 4 mm)	Eluent A: ACN Eluent B: H_2O	0.7/Gradient/NI	DAD (λ: 224 nm)	37-41 µg/kg	[124]
Trichothecenes A	LiChrospher100 C-18 (250 x 4 mm)	ACN/H_2O, 0.75 % formic acid (65:35)	1.0/Isocratic/NI	FL (pre-column derivatization with coumarin-3-carbonyl chloride, (λ_{ex}: 292 nm; λ_{em}: 425 nm)	10-15 µg/kg	[125]
Trichothecenes A, B	Spherisorb S3ODS2 (250 x 2.1 mm)	Eluent A: Methanol Eluent B: H_2O	1.0/Gradient/NI	FL (pre-colum derivatization with coumarin-3-carbonyl chloride, λ_{ex}: 292 nm; λ_{em}: 425 nm)	0.2-1.0 µg/kg	[126]

Analyte	Column	Eluent	Flow/Mode/Temp	Detection	LOD	Ref.
Trichothecenes	Lichrospher 100 C$_{18}$ (250 x 4 mm)	ACN/H$_2$O (65:35)	1.0/Isocratic/RT	FL (pre-column derivatizion with coumarin-3-carbonyl chloride, λ_{ex}: 292 nm; λ_{em}: 425 nm)	10 μg/kg	[127]
Ochratoxin A, zearalenone	Luna Phenyl Hexyl (250 x 4.6 mm)	Eluent A: 0.25 M H$_3$PO$_4$ Eluente B: Methanol:H$_2$O (9:1)	0.6/Isocratic/40	FL (Ochratoxin A: λ_{ex}: 332 nm; λ_{em}: 500 nm; Zearalenone: λ_{ex}: 278 nm; λ_{em}: 460 nm)	Ochratoxin: 0.1 μg/kg; Zearalenone: 1.0 μg/kg	[80]
Zearalenone	Eurospher 100 RP-18 (250 x 4 mm)	Eluent A: ACN Eluent B: H$_2$O	1.0/Gradient/NI	FL (λ_{ex}: 271 nm; λ_{em}: 452 nm)	4 μg/kg	[124]
	Lichrospher 100 C18 (250 x 4 mm)	Methanol/H$_2$O (65:35)	1.0/Isocratic/RT	FL (λ_{ex}: 274 nm; λ_{em}: 440 nm)	10 μg/kg	[127]
	XBridge C18 (250 x 4.6 mm)	Eluent A: H$_2$O /ACN/HClO$_4$ (97:3:0.01) Eluent B: ACN/ HClO$_4$ (100:0.01)	1.0/Gradient/15	DAD (λ: 280 nm)	0.09 μg/L	[128]
Patulin	Alltech Hypersil BDS C18 (250 x 4.6 mm)	Eluent A: ACN Eluent B: H$_2$O	NI/Gradient/NI	DAD (λ: 276 nm)	25 μg/L	[130]

Table 5. (Continued)

Phenomenex Synergi Polar RP80A (250 x 4.6 mm)	ACN/H$_2$O (1.5:98.5)	1.0/Isocratic/RT	DAD (λ: 275 nm)	10 µg/kg	[131]
Shim-pack CLC-ODS (200 x 4.6 mm)	Eluent A: H$_2$O Eluent B: 30 % ethanol in water	0.5/Gradient/NI	DAD (λ: 276 nm)	4 µg/L	[132]

[1] Abbreviations: ACN: Acetonitrile; DAD: Diode array detection; ELSD: Evaporative laser scattering detector; FL: Fluorescence; NI: Not indicated; NDA: Naphthalene-2,3-dicarboxaldehyde; OPA: *o*-phtalaldehyde; PBPB: Pyridinium hydrobromide perbromide; RT: Room Temperature; TFA: Trifluoroacetic acid.

as the need to derivatize the samples before analysis have led to a prevalence of LC-MS over GC-MS approaches [77]. Recently, LC-MS methods have been reviewed by Sforza et al [3] and by Zöllner et al [19], and it has been showed that MS technique will become a robust method for multi-component mycotoxin analysis with more accurate quantification, better selection and higher sensitivity compared to other chromatographic methods.

Several analyzers can be used for the detection of mycotoxins. Single quadrupole provides sufficient sensitivity in SIM mode [87], although ion trap [134], triple quadrupole [135], quadrupole-ion trap (Q-TRAP) [136] or TOF [137] can be also used, offering higher confidence for the quantification of the compounds.

As it can be observed in Table 6, ochratoxin A can be analyzed by LC-MS using several analyzers such as ion trap [68], Q-TRAP [67] or triple quadrupole [88]. This mycotoxin can be detected in positive or negative mode. In positive mode, it can form either proton adduct at *m/z* 404 or sodium adduct at *m/z* 426. However, using methanol and water as mobile phases and a relatively low concentration of formic acid as mobile phase modifier, negative ionization can give a more intense ion at *m/z* 402 ([M-H]$^-$), bearing in mind that generally, in ESI negative, the responses of the compounds containing a carboxylic acid are decreased as formic acid concentration increases. On the other hand, the presence of acid helps to retain analytes on the column and gives a better chromatographic resolution and lower matrix effect [67].

In the same way, aflatoxins can be either analyzed by positive [138] or negative mode [73] as well as zearalenone [87,139], whereas patulin is usually analyzed in negative mode [140], selecting *m/z* 153 as quantification ion. When MS is used, it allows the reduction of the influence of matrix interferences during the quantification step, using electrospray (ESI) [141], atmospheric pressure chemical ionization (APCI) [142] or atmospheric pressure photoionization (APPI) [143] as ionization sources.

On the other hand, fumonisins are usually analyzed using positive ionization [144,145], improving the possibilities of employing LC-MS in the analysis of these type of compounds [146]. For these mycotoxins, the mobile phase is usually acidified with formic acid to obtain symmetrical peak shapes due to the presence of four carboxylic groups in their molecular structure [45].

For the analysis of trichothecenes, positive or negative ionization can be used, applying ESI or APCI [147]. For instance, negatively charged molecular ions of type B trichothecenes (nivalenol, deoxynivalenol) were detected by far more sensitively than their positively charged analogues [42]. However, trichothecenes type A provides better sensitivity in positive mode [148], giving the pseudomolecular ion [M-H]$^+$, so the development of a method for simultaneous determination of trichothecenes, polarity switching during chromatography should be carried out [149]. In this case, the use of a buffer such as ammonium formate or acetate leads to the formation of single trichothecene-adduct ion species, and the intensities were much higher compared to experiments without buffers. Using ammonium acetate, the peak areas of both [M-NH$_4$]$^+$ ions for trichothecenes type A in the positive ionization mode and [M-CH$_3$COO]$^-$ ions for trichothecenes type B in the negative ionization mode, increases the sensitivity of the method [77].

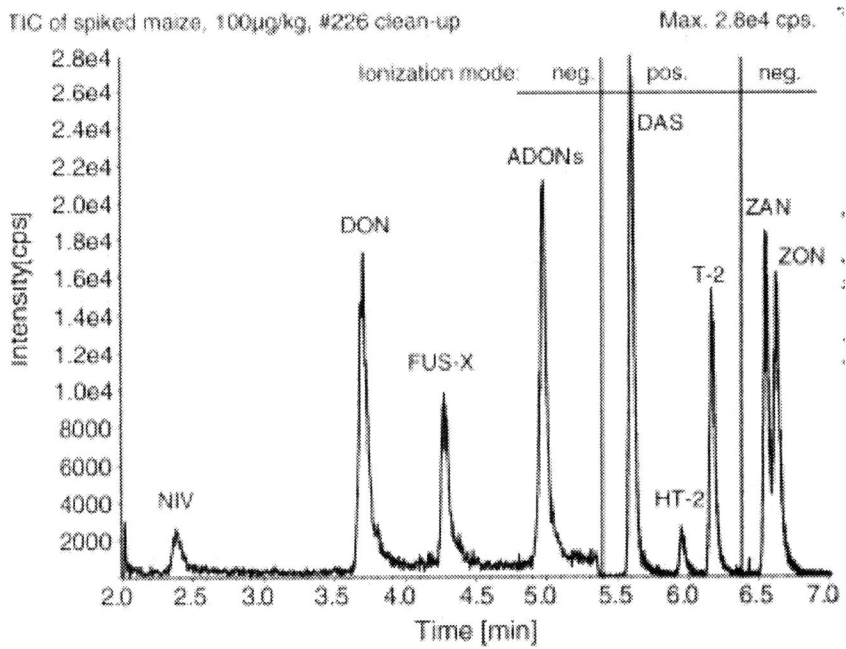

Figure 5. Total ion chromatogram obtained after clean-up with MycoSep 226 columns and LC–MS/MS analysis of a spiked maize sample containing 100 µg/kg of each mycotoxin. Vertical lines illustrate change of ionization polarities from negative to positive (5.4 min) and back to negative (6.4min). Abbreviations: ADONs: 3-acetyl and 15-acetyl deoxynivalenol; DAS: diacetoxyscirpenol; DON: deoxynivalenol; FUS-X: fusarenon X; HT-2: HT-2 toxin; NIV: T-2: T-2 toxin; ZAN:zearalanone (internal standard); ZON: zearalenone. Reprinted from [77], copyright 2006, with permission from Elsevier.

This approach can be extended for the methods developed for the multi-mycotoxins analysis, which have been taking into account that for the determination of mycotoxins having opposed polarity, it has to be accepted that certain conditions may be far from optimal for some of the analytes. In order to increase the sensitivity during MS detection, the mobile phase for mycotoxins analyzed in positive ion mode can be acidified with acetic or formic acid. However, acidified mobile phase could not be used for mycotoxins analyzed in negative ion mode because it results in poor signal intensities. Therefore the addition of ammonium acetate or formate can increase the signal intensity significantly [82].

In Figure 5 it can be observed a chromatogram with polarity switching that allows the simultaneous determination of trichothecenes A, B and zearalenone, selecting the optimum detection conditions for each group of mycotoxins.

In relation to the organic solvent used in the mobile phase, methanol and acetonitrile can be selected. In general, if acetonitrile was chosen as the mobile phase, the ionization of the mycotoxins was reduced, whereas if methanol was selected, higher sensitivity can be observed [92].

One of the main disadvantages of LC-MS/MS is the influence of co-extracted matrix compounds on the signal response of target molecules (matrix effects). Stable isotope-labeled reference compounds can be applied as internal standards (IS) within quantitative mass spectrometry, as it can be observed in Figure 6, where $^{13}C_3$-labelled patulin was used for the determination of patulin in fruit juices.

However there are not commercially available isotopically labelled compounds for all the mycotoxins and other strategies such as matrix-matched calibration or standard addition methodology can be applied [92]. In this sense, if matrix effect is used, it can be detected some differences in the matrix effect within a given matrix [150], and standard addition methodology can be used when mycotoxin concentrations were close to the limits indicated in international regulations in order to assure the reliability of the obtained results.

Finally, it must be mentioned that in the last few years, ultra high performance liquid chromatography (UHPLC) has been coupled to MS in order to reduce analysis time, as well as it increases sensitivity, resolution and speed compared to LC. These advantages are based on the use of columns filled with particle size lower than 2 µm and instruments with high pressure fluidic modules. Up to know few works have been used this approach [55,82,93], although in the near future this will be implemented in routine laboratories, taking into account that mycotoxins can be analyzed in less than 10 minutes, increasing sample throughput. For instance, Figure 7 shows a typical chromatogram of a blank maize spiked with twelve mycotoxins [93], observing that the separation of the target compounds was achieved in less than 4 minutes. Although complete resolution was not obtained, the use of MS/MS detection allows the selective analysis of all the compounds.

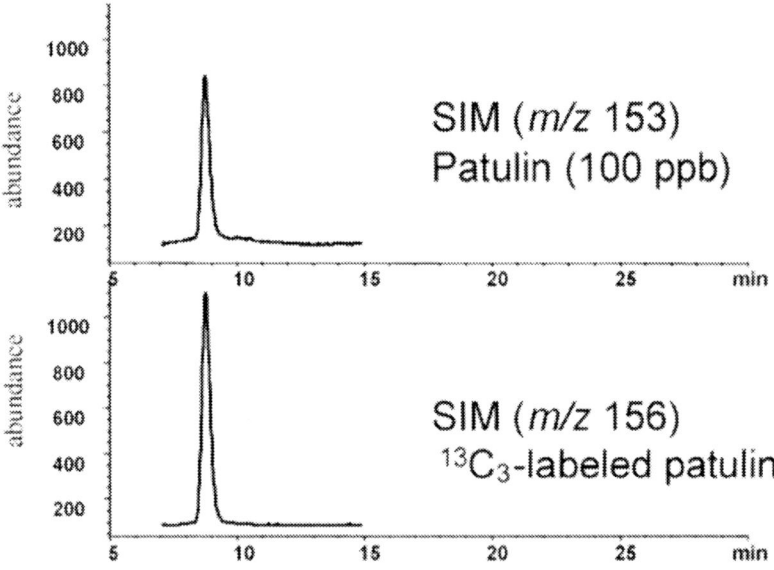

Figure 6. Selected ion monitoring chromatograms of patulin (m/z 153) and 13C3-labeled patulin as internal standard (m/z 156). The chromatogram was obtained with 100 ppb of patulin standard, and 0.5 ng of patulin was injected. Reprinted from [48], copyright 2004, with permission from American Chemical Society.

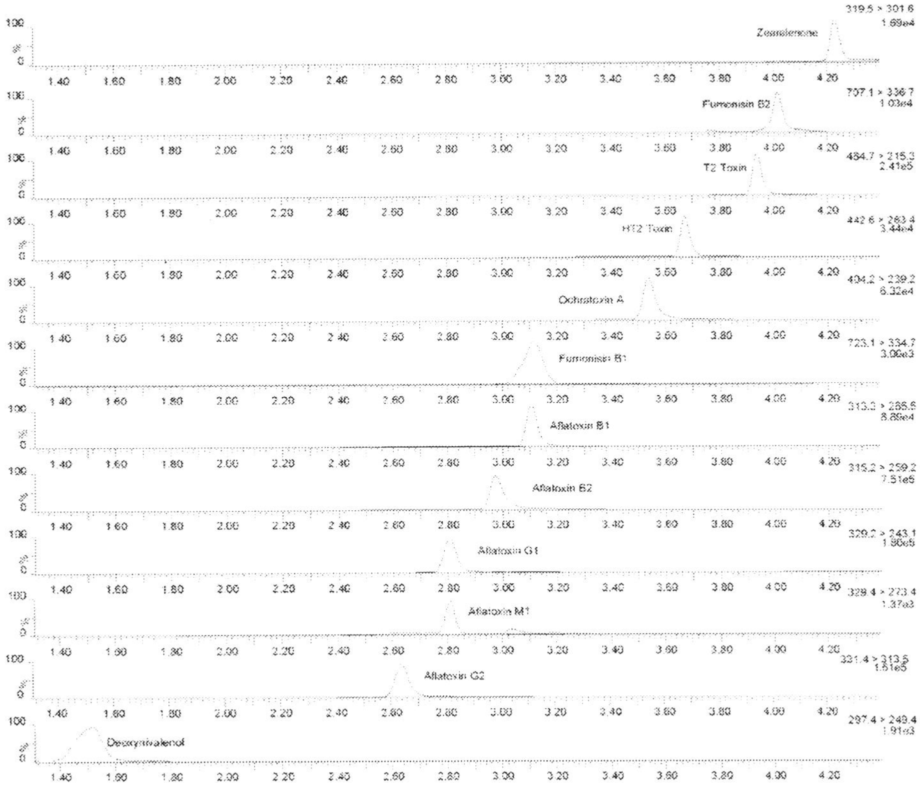

Figure 7. UHPLC-MS/MS chromatograms obtained from a blank maize sample spiked at 25 μg/kg (2.5 μg/kg for aflatoxin M1). Reprinted from [93], copyright 2009, with permission from Elsevier.

Table 6. Liquid chromatographic methods for the determination of mycotoxins in foodstuffs using MS detection1

Mycotoxin	Stationary phase	Mobile phase	Flow (mL/min)/Mode/Column temperature (°C)	Detection conditions	Detection limit	Ref
Ochratoxin A						
	Inertsil ODS 3 (250 x 2 mm)	Methanol/H_2O/acetic acid (70:30:1.5)	0.25/Isocratic/NI	IT (MS/MS) 404 > 358 (ESI, positive ionization)	1.7 µg/kg	[68]
	Luna C_{18} (150 x 4.6 mm)	ACN/H_2O/acetic acid (99:99:2)	0.7/Isocratic/NI	IT (MS/MS) ESI (positive ionization)	0.01 µg/L	[134]
	Alltima C_{18} (150 x 1 mm)	Eluent A: H_2O (10 mM formic acid). Eluent B: Methanol (10 mM formic acid)	1.0/Gradient/NI	QTRAP (MS/MS) 402 > 358 (ESI, negative ionization)	0.01 µg/L	[67]
	Alltima C_{18} (150 x 3.2 mm)	Eluent A: ACN (0.3 % Formic acid) Eluent B: H_2O (0.3 % formic acid)	0.3/Gradient/NI	QqQ (MS/MS) 404 > 239 and 404 > 341 (ESI positive ionization)	0.11 µg/kg	[88]

Analyte	Column	Eluent	Flow/Mode/Temp	Detection	LOD	Ref
Aflatoxin M$_1$	PRP-1 poly(styrene-divinilbenzo, SDB) (100 x 2.1 mm)	Eluent A: 10 mM 4-metnylmorpholine (pH 9.7) Eluent B: ACN	0.2/Gradient/40	QqQ (MRM) 327 > 312 (ESI, negative ionization)	0.6 ng/L	[73]
Aflatoxins	LiChrocart C$_{18}$ (30 x 4 mm)	Methanol/H$_2$O (30:70)	1.0/Isocratic/30	Q(SIM) ESI (positive ionization)	10 ng	[138]
Patulin	Mightysil RP-18 GP (250 x 2 mm)	Eluent A: H$_2$O Eluent B: ACN	0.2/Gradient/NI	Q (SIM) m/z 153 (ESI, negative ionization)	2.5 pg in sample	[48]
	Synergi MAX RP 80 A (150 x 4.6 mm)	ACN/ H$_2$O (20:80)	0.75/Isocratic/40	Q (SIM) m/z 153 (ESI, negative ionization)	23.5 ng/L	[141]
	Zorbax Eclipse XDB C18 (12.5 x 4.6 mm)	Methanol/10 mM ammonium acetate (2:98)	0.3/Isocratic/30	Q (SIM) m/z 153 (APPI, negative ionization)	1.5 µg/L	[143]
	Nucleosil RP18 (250 x 2 mm)	Eluent A: ACN Eluent B: H$_2$O	0.4/Gradient/NI	IT (MS) ESI (negative ionization) IT (MS/MS)	20.3 µg/L	[140]
	Luna C$_{18}$ (150 x 4.6 mm)	ACN/H$_2$O (10:90)	0.5/Isocratic/NI	APCI (positive and negative ionization)	4 µg/L	[142]

Table 6. (Continued)

Analyte	Column	Eluent	Flow/Gradient	Detection	LOD	Ref.
Fumonisin B_1	J'Sphere ODSL 18 (150 x 4.6 mm)	Eluent A: H_2O (0.02 % formic acid) Eluent B: ACN (0.02 % formic acid) Eluent A: 0.1 % TFA	1.0/Gradient/NI	QqQ and IT (MS/MS) ESI (positive ionization)	NI	[144]
Fumonisin B_1, B_2	Pecosphere C_{18} (33 x 4.6 mm)	Eluent B: Methanol/5 mM ammonium acetate (80:20) with 1 % formic acid	0.03/Gradient/NI	Q (SIM) ESI (positive ionization)	5 µg/kg	[146]
Fumonisins	Ultracarb 3 ODS C_{18} (150 x 2mm)	ACN/0.1 % formic acid (45:55)	0.2/Isocratic/NI	Q (SIM) ESI (positive ionization)	5 ng	[145]
	Supelcosil ABZ Plus (250 x 2.1 mm)	Eluent A: H_2O (0.1 % formic acid) Eluent B: ACN (0.1 % formic acid)	0.3/Gradient/40	IT (MS/MS) ESI (positive ionization)	NI	[45]
Zearalenone	Hypersil ODS (100 x 2.1 mm)	ACN/H_2O (40:60)	1.0/Gradient/35	Q (SIM) m/z 319 (APCI, positive ionization)	0.12 µg/kg	[87]
	Supersphere 100 RP-18 (125 x 3 mm)	Methanol/H_2O (15 mM ammonium acetate, pH 7.5)	0.5/Isocratic/NI	QqQ (MS/MS) 317 > 131 and 317 > 175 (APCI; negative ionization)	0.5 µg/kg	[135]

Analyte	Column	Eluent	Flow/Mode/Time	Detection	Range	Ref
Deoxynivalenol	Aquasil RP-18 (100 x 4.6 mm)	Methanol/H$_2$O (15:85)	0.8/Isocratic/22	QTRAP (MS/MS) 295.3 > 265.3 APCI (negative ionization)	6 μg/kg	[136]
Deoxynivalenol, zearalenone	Jones C$_{18}$ (150 x 2.1 mm)	Eluent A: H$_2$O Eluent B: Methanol	0.2/Gradient/30	QqQ (MS/MS) ESI (negative ionization)	0.01-0.1 μg/kg	[41]
T2-toxin, deoxynivalenol	SB-RP18-Zorbax (150 x 3 mm)	Eluent A: Methanol Eluent B: H$_2$O	0.25/Gradient/NI	QqQ (MS/MS) ESI (positive and negative ionization)	0.01-17 μg/kg	[42]
Trichothecenes A	Hypersil C$_{18}$ (200 x 2.1 mm)	Eluent A: 1 mM Ammonium acetate Eluent B: AcN	0.3/Gradient/30	Q (SIM) APCI (positive ionization)	50-85 μg/kg	[148]
Trichothecenes A,B	Synergi Fusion RP 80 A (250 x 2 mm)	Eluent A: 0.00184 mM ammonia and 0.13 mM ammonium acetate Eluent B: can Eluent A: ACN (0.07 mM ammonium acetate)	0.2/Gradient/25	QqQ (MS/MS) ESI (positive and negative ionization)	0.18-5 μg/kg	[149]
Trichothecenes A, B, D	Alltima C$_{18}$ (250 x 4.6 mm)	Eluent B: H$_2$O (0.07 mM ammonium acetate)	0.15/Gradient/NI	QqQ (MS/MS) ESI (positive ionization)	0.03-50 μg/kg^2	[40]
Trichothecenes	Alltima C$_{18}$ (250 x 4.6 mm)	Eluent A: ACN/Methanol (70:30) Eluent B: H$_2$O	1.0/Gradient/NI	QqQ (MS/MS) APCI (negative ionization)	1.8-10 μg/kg	[39]

Table 6. (Continued)

	Nucleosil C$_{18}$ (125 x 2 mm)	Eluent A: Methanol Eluent B: H$_2$O	0.25/Gradient/Nl	IT (MS/MS) APCI (positive ionization)	1-10 µg/kg	[147]
Trichothecenes A, B and zearalenone	Aquasil RP-18 (100 x 4.6 mm)	Eluent A: Methanol/H$_2$O (5 mm ammonium acetate, 20:80) Eluent B: Methanol/H$_2$O (5 mm ammonium acetate, 90:10)	1.0/Gradient/25	Q-TRAP (MS/MS) APCI (positive and negative ionization)	0.3-3.8 µg/kg	[77]
Trichothecenes, aflatoxins and zearalenone	Zorbax Eclipse XDB C$_{18}$ (150 x 2.1 mm)	Eluent A: H$_2$O (10 mM ammonium acetate) Eluent B: Methanol	0.2/Gradient/40	TOF (MS) APCI (positive ionization)	0.1-6.1 µg/kg	[86]
Trichothecenes, fumonisins and zearalenone	Alltima C$_{18}$ (250 x 2.1 mm)	Eluent A: Methanol (10 mM formic acid) Eluent B: H$_2$O (10 mM formic acid).	0.2/Gradient/45	QqQ (MS/MS) ESI (positive and negative ionization)	2-14 µg/kg	[89]
Trichothecenes, fumonisins, macrocyclic lactones	Alltima C$_{18}$ (250 x 2.1 mm)	Eluent A: H$_2$O (0.1 % formic acid) Eluent B: Methanol (0.1 % formic acid)	0.2/Gradient/45	QqQ (MS/MS) ESI (positive and negative ionization)	1-19 µg/kg^2	[43]
12 mycotoxins	UPLC BEH	Eluent A: Methanol	0.35/Gradient/30	QqQ (MS/MS)	0.01-2.1	[93]

16 mycotoxins	C_{18} (100 x 2.1 mm)	Eluent B: H_2O (5 mM ammonium formate)		ESI (positive ionization)	µg/kg	
	Sunfire C_{18} (150 x 3 mm)	Eluent A: H_2O Eluent B: Methanol	0.3/Gradient/Nl	QqQ (MS/MS) ESI (positive ionization)	0.04-74 µg/L	[81]
17 mycotoxins	UPLC BEH C_{18} (100 x 2.1 mm)	Eluent A: 10 mM ammonium acetate Eluent B: Methanol Eluent A:	0.3/Gradient/35	QqQ (MS/MS) ESI (positive and negative ionization)	0.01-0.70 µg/kg	[82]
18 mycotoxins	Luna C_{18} 100A (150 x 4.6 mm)	Methanol/H_2O (10:90) Eluent B: Methanol/H_2O (90:10)	0.2/Gradient/25	QqQ (MS/MS) ESI (positive and negative ionization)	0.02-0.15 µg/L	[37]
33 mycotoxins	Alltima C_{18} (150 x 3.2 mm)	Eluent A: H_2O (0.1 % formic acid) Eluent B: ACN (0.1 % formic acid) Eluent A:	0.3/Gradient/30	QqQ (MS/MS) ESI (positive ionization)	0.5-200 µg/kg	[78]
39 mycotoxins	Gemini C18 (150 x 4.6 mm)	Methanol/H_2O/acetic acid (10:89:1) Eluent B: Methanol/H_2O/acetic acid (97:2:1)	1.0/Gradient/25	QqQ (MS/MS) ESI (positive and negative ionization)	0.03-220 µg/kg	[83]

Table 6. (Continued)

87 mycotoxins	Gemini C$_{18}$ (150 x 4.6 mm)	Eluent A: Methanol/H$_2$O/acetic acid (10:89:1) Eluent B: Methanol/H$_2$O/acetic acid (97:2:1)	1.0/Gradient/25	QTRAP (MS/MS) ESI (positive and negative ionization)	0.02-225 µg/kg	[92]
474 mycotoxins and fungal metabolites	Hypersil BDS-C$_{18}$ (125 x 2 mm)	Eluent A: H$_2$O Eluent B: ACN	0.3/Gradient/NI	DAD-TOF ESI (positive ionization)	1 pg- 3 ng	[137]

[1] Abbreviations: ACN: Acetonitrile APCI: Atmospheric pressure chemical ionization; APPI: Atmospheric pressure photoionization; DAD: Diode array detection; ESI: Electrospray; IT: Ion trap; NI: Not indicated; Q: Single quadrupole; QqQ: Triple quadrupole; SIM: Selected ion monitoring; TOF: Time of flight.

[2] Quantification limit.

Chapter 5

CONCLUSION

Nowadays, there is no doubt in relation to the use of MS for the unambiguous determination of mycotoxins. LC-MS applications in mycotoxin analyses include multi-methods that are able to detect a high amount of these compounds. However, one of the main problems is the quantification step due to the presence of matrix effects. The use of labeled internal standards is becoming very popular, although many of these standards are quite expensive, and alternative approaches such as matrix-matched calibration or standard addition methodology must be used.

Despite of the use of chromatographic methods for the determination of mycotoxins, the potential of LC-MS/MS for screening and high-throughput multi-toxin analysis has recently been demonstrated, indicating that this approach will have a strong presence in future trends in mycotoxin analysis [96].

Several analyzers can be used such as single or triple quadrupole or ion traps. However, other analyzers such as TOF are scarcely used although it offers several advantages. For instance, TOF has the capability, in contrast to quadrupole or ion trap instruments, of detecting a wide mass range without losing significant sensitivity. Furthermore, TOF instruments provide enhanced full mass range spectra sensitivity and accuracy due to their higher mass resolution. This also makes it possible to distinguish among isobaric ions and increases confidence in the identification of analytes by giving an estimate of the elemental composition of each ion. The information obtained from TOF analysis has the additional advantage that quantification can be performed on any ion observed in the acquired mass range. These properties should be of great advantage for the screening of foodstuffs containing multiple mycotoxins with different molecular weights. Furthermore, it can be used for the identification of unknown compounds present at low concentrations and for suggesting their structures. The detection of new compounds and their precursors may yield further data for a more detailed knowledge of their biosynthesis. Despite of TOF, the combination of several analyzers can provide more details regarding structural data of these new compounds [45].

Most of the developed methods are focused on the determination of aflatoxins, fumonisins, trichothecenes, ochratoxin A, and patulin, but spoilage fungi such as *Aspergillus*, *Penicillium* or *Alternaria* can produce remaining mycotoxins that are able to infect food and feed upon storage. For these mycotoxins, there is a lack of comprehensive data on their presence in spoiled food, which makes it difficult to evaluate the health hazard that they pose to the end consumer, and new chromatographic methods must be developed in order to determine these compounds. In this sense, the combination of LC or UHPLC with high resolution mass spectrometry analyzers will be a valuable tool for the identification and quantification of these type of compounds.

Although there have been several recent successes in detection of mycotoxins, new methods are still required to achieve higher sensitivity and address other challenges that are posed by these toxins. The application of MS in conjunction with other tools for decreasing limits of detection has been of increased interest. Future trends would focus on rapid assays and tools that would measure multiple toxins from a single matrix.

ACKNOWLEDGMENTS

The authors gratefully acknowledge Spanish Ministry of Education and Science (MEC-FEDER) (Project Ref. AGL2006-12127-C02-01) for financial support. RRG is also grateful for personal funding through the Ramon y Cajal program (Spanish Ministry of Education and Science-EFS).

REFERENCES

[1] WHO/FAO. Safety Evaluation of Certain Mycotoxins in Food. WHO Food Additives Series 47/FAO Food and Nutrition Paper 74, WHO: Geneva, 2001.
[2] Shephard, G. S. *Chem. Soc. Rev.* 2008, *37*, 2468-2477.
[3] Sforza, S.; Dall'Asta, C.; Marchelli, R. *Mass Spectrom. Rev.* 2006, *25*, 54-76.
[4] Miller, J. D. *Food Addit. Contam.* 2008, *25*, 219-230.
[5] Bucheli, P.; Taniwaki, H. *Food Addit. Contam.* 2002, *19*, 655-665.
[6] Drusch, S.; Ragab, W. *J. Food Protect.* 2003, *66*, 1514-1527.
[7] Coffey, R.; Cummins, E.; Ward, S. *Food Control* 2009, *20*, 239-249.
[8] Ventura, M.; Guillén, D.; Anaya, I.; Broto-Puig, F.; Lliberia, J. L.; Agut, M.; Comellas, L. *Rapid Commun. Mass Spectrom.* 2006, *20*, 3199-3204.
[9] Ribeiro, E.; Alves, A. *Anal. Bioanal. Chem.* 2008, *391*, 1443-1450.
[10] Kuiper-Goodman, T. In *Mycotoxins and Phycotoxins. Developments in Chemistry, Toxicology and Food Safety,* Miraglia M, van Edmond H, Brera C.; Alaken Inc.: Fort Collins, CO, 1998; pp 25.
[11] [11] Commission Regulation (EC) No 1881/2006, 2006. Commission Directive 2006/1881/EC of 19 December 2006, setting maximum levels for certain contaminants in food stuffs. 2006. *Official Journal of the European Communities*, L364, 5-24.
[12] FAO, Worldwide regulations for mycotoxins in food and feed in 2003. FAO Food and Nutrition Paper No. 81. Rome, Italy, 2006.
[13] Krska, R.; Molinelli, A. *Anal. Bional. Chem.* 2009, *393*, 67-71.
[14] Turner, N. W.; Subrahmanyam, S.; Piletsky, S. A. *Anal. Chim. Acta* 2009, *632*, 168-180.
[15] Papp, E.; Otta, K.; Záray, G.; Mincsovics, E. *Microchem. J.* 2002, *73*, 39-46.

[16] Lin, L.; Zhang, J.; Wang, P.; Wang, Y.; Chen, J. *J. Chromatogr. A* 1998, *815*, 3-20.
[17] Gilbert, J.; Anklam, E. *Trends Anal. Chem.* 2002, *21*, 468-486.
[18] Jaimez, J.; Fente, C. A.; Vázquez, B. I.; Franco, C. M.; Cepeda, A.; Mahuzier, G.; Prognon, P. *J. Chromatogr. A* 2000, *882*, 1-10.
[19] Zöllner, P.; Mayer-Helm, B. *J.Chromatogr. A* 2006, *1136*, 123-169.
[20] Comission Decision 2002/657/EC of 12 August 2002 implementing Council Directive 96/23/EC concerning the performance of analytical methods and the interpretation of results, *Official Journal of the European Communities*, L221, 2002, 8-36.
[21] Lo Curto, R.; Pellicano, T.; Vilasi, F.; Munafo, P.; Dugo, D. *Food Chem.* 2004, *84*, 71-75.
[22] Peraica, M.; Radic, B.; Lucic, A.; Pavlovic, M. *Bull. WHO* 1999, *77*, 754-766.
[23] Amezqueta, S.; González-Peñas, E.; Murillo-Arbizu, M.; López de Cerain, A. *Food Control* 2009, *20*, 326-333.
[24] Ringot, D.; Chango, B.; Schneider Y. J.; Larondelle, Y. *Chem-Biol Interact.* 2006, *159*, 18-46.
[25] Serra, R.; Abrunhosa, L.; Kozakiewicz, Z.; Venancio A. *Int. J. Food Microbiol.* 2003, *88*, 63-68.
[26] Serra, R.; Mendonca, C.; Abrunhosa, L.; Pietro, A. Venancio, A. *Anal. Chim. Acta* 2004, *513*, 41-47.
[27] Battilani, P.; Pietri, A. *Eur. J. Plant Pathol.* 2002, *108*, 639-643.
[28] López de Cerain, A.; González-Peñas, E.; Jiménez, A. M.; Bello, J. *Food Addit. Contam.* 2002, *19*, 1058-1064.
[29] Codex Alimentarius Comisión, Position paper on Ochratoxin A, 1998, CX/FAC/99/14.
[30] Taniwaki, M.; Pitt, J.; Teixeira, A.; Iamanaka, B. *Int. J. Food Micr.* 2003, *82*, 173-179.
[31] van Barneveld, R. J. *Aust. J. Agric. Res.* 1999, *50*, 807-823.
[32] Castegnaro, M.; Tozlovanu, M.; Wild, C.; Molinie, A.; Sylia, A.; Pfohl-Leszkowicz, A. *Mol. Nutr. Food Res.* 2006, *50*, 480-487.
[33] Lu, F. C. *Environ. Health Prev. Med.* 2003, *7*, 235-238.
[34] Anklam, E.; Stroka, J.; Boenke, A. *Food Control* 2002, *13*, 173-183.
[35] Kuilman, M. E. M.; Mass, R. F. M.; Judah, D. J.; Fink-Gremmels, J. *J. Agric. Food Chem.* 1998, *46*, 2707-2713.
[36] Hussein, H. S. Brasel, J. M. *Toxicology* 2001, *167*, 101-134.
[37] Sorensen, L. K.; Elbaek, T. H. *J. Chromatogr. B* 2005, *820*, 183-196.

[38] Manetta, A. C.; Guiseppe, L.; Giammarco, M.; Fusaro, I.; Simonella, A.; Gramenzi, A.; Formigoni, A. *J. Chromatogr. A* 2005, *1083*, 219-222.
[39] Laganá, A.; Curini, R.; D'Ascenzo, G.; De Leva I.; Faberi, A.; Pastorini, E. *Rapid Commun. Mass Spectrom.* 2003, *17*, 1037-1043.
[40] Gentili, A.; Caretti, F.; D'Ascenzo, G.; Mainero-Rocca, L.; Marchese, S.; Materazzi, S.; Perret, D. *Chromatographia* 2007, *66*, 669-676.
[41] Sypecka, Z.; Kelly, M.; Brereton, P. *J. Agric. Food Chem.* 2004, *52*, 5463-5471.
[42] Biselli, S.; Hummert, C. *Food Addit. Contam.* 2005, *22*, 752-760.
[43] Cavaliere, C.; Foglia, P.; Guarino, C.; Motto, M.; Nazzari, M.; Samperi, R.; Laganá, A.; Berardo, N. *Food Chem.* 2007, *105*, 700-710.
[44] Shephard, G. S. *J. Chromatogr. A* 1998, *815*, 31-39
[45] Bartok, T.; Szecsi, A.; Szekeres, A.; Mesterhazy, A.; Bartok, M. *Rapid Commun. Mass Spectrom.* 2006, *20*, 2447-2462.
[46] Chu, F. S.; Li, G. Y. *Appl. Environ. Microbiol.* 1994, *60*, 847-852.
[47] Sewram, V.; Nair, J. J.; Nieuwoudt, T. W.; Leggott, N. L.; Shephard, G. S. *J. Chromatogr. A* 2000, *897*, 365-374.
[48] Ito, R.; Yamazaki, H.; Inoue, K.; Yoshimura, Y.; Kawaguchi, M.; Nakazawa, H. *J. Agric. Food Chem.* 2004, *52*, 7464-7468.
[49] Jiménez, M.; Máñez, M.; Hernández, E. *Int. J. Food Microbiol.* 1996, *29*, 417-421.
[50] Krska, R.; Schubert-Ullrich, P.; Molinelli, A.; Sulyok, M.; MacDonald, S.; Crews, C. *Food Addit. Contam.* 2008, *25*, 152-163.
[51] Commission Regulation (EC) No 1126/2007, 2007. Commission Regulation 2007/1126/EC of 28 September 2007 amending Regulation (EC) No 1881/2006 setting maximum levels for certain contaminants in foodstuffs as regards Fusarium toxins in maize and maize products. *Official Journal of the European Communities*, L255, 14-17.
[52] JEFCA (Joint FAO/WHO Expert Comité on Food Additives). Ochratoxin A. In *Safety Evaluation of Certain Mycotoxins in Food.* WHO Food Additives Series 47; FAO Food and Nutrition Paper 74; WHO: Geneva, Switzerland, 2001; p 366.
[53] World Health Organization. Evaluation of certain food additives and contaminants. *44th Report of the Joint FAO/WHO Expert Committee on Food Additives*; Technical Report Series 859; Geneva, Switzerland, 1995; pp 36-38.
[54] Van Egmond, H. P.; Schothorst, R. C.; Jonker, M. A. *Anal. Bional. Chem.* 2007, *389*, 147-157.

[55] Mol, H. G. J.; Plaza-Bolaños, P.; Zomer, P.; De Rijk, T. C.; Stolker, A. A. M.; Mulder, P. P. J. *Anal. Chem.* 2008, *80*, 9450-9459.
[56] Medina, A.; Jiménez, M.; Gimeno-Adelantado, J. V.; Valle-Algarra, F. M.; Mateo, R. *J. Chromatogr. A* 2005, *1083*, 7-13.
[57] Aresta, A.; Vatinno, R.; Palmesano, F.; Zambonin, C. G. *J. Chromatogr. A* 2006, *1115*, 196-201.
[58] González-Peñas, E.; Leache, C.; Viscarret, M.; Pérez de Obanos, A.; Araguas, C.; López de Cerain, A. *J. Chromatogr. A* 2004, *1025*, 163-168.
[59] Zimmerli, B.; Dick, R. *J. Chromatogr. B* 1995, *666*, 85-99.
[60] Visconti, A.; Pascale, M.; Centonze, G. *J. Chromatogr. A* 1999, *864*, 89-101.
[61] Visconti, A.; Pascale, M.; Centonze, G. *J. Chromatogr. A* 2000, *888*, 321-326.
[62] Blesa, J.; Soriano, J. M.; Moltó, J. C.; Mañés, J. *J. Chromatogr. A* 2004, *1054*, 397-401.
[63] Leitner, A.; Zöllner, P.; Paolillo, A.; Stroka, J.; Papadopoulou-Bouraoui, A.; Jaborek, S.; Anklam, E.; Lindner, W. *Anal. Chim. Acta* 2002, *453*, 33-41.
[64] Monaci, L.; Palmesano, F. *Anal. Bioanl .Chem.* 2004, *378*, 96-103.
[65] Brera, C.; Grossi, S.; De Santis, B.; Miraglia, M. *J. Liq. Chromatogr. Relat. Technol.* 2003, *26*, 119-133.
[66] Sáez, J. M.; Medina, A.; Gimeno-Adelantado, J. V.; Mateo, R.; Jiménez, M. *J. Chromatogr. A* 2004, *1029*, 125-133.
[67] Bacaloni, A.; Cavaliere, C.; Faberi, A.; Pastorini, E.; Samperi, R.; Laganá, A. *J. Agric. Food Chem.* 2005, *53*, 5518-5525.
[68] Reinsch, M.; Töpfer, A.; Lehmann, A.; Nehls, I. *Anal. Bional. Chem.* 2005, *381*, 1592-1595.
[69] Murillo-Arbizu, M.; Amezqueta, S.; Gónzalez-Peñas, E.; López de Cerain, A. *Food Chem.* 2009, *113*, 420-423.
[70] Moukas, A.; Panagiotopoulou, V.; Markaki, P. *Food Chem.* 2008, *109*, 860-867.
[71] Li, J.; Wu, R.; Hu, Q.; Wang, J. *Food Control* 2007, *18*, 530-534.
[72] Gökmen, V.; Acar, J.; Sarioglu, K. *Anal. Chim. Acta* 2005, *543*, 64-69.
[73] Chen, C. Y.; Li, W. J.; Peng, K. P. *J. Agric. Food Chem.* 2005, *53*, 8474-8480.
[74] Muscarella, M.; Lo Magro, S.; Palermo, C.; Centonze, D. *Anal. Chim. Acta* 2007, *594*, 257-264.
[75] Shundo, L.; Navas, S.A.; Conceicao, L.; Lamardo, A.; Ruvieri, V.; Sabino, M. *Food Control* 2009, *20*, 655-657.

[76] Cavaliere, C.; Foglia, P.; Guarino, C.; Nazzari, M.; Samperi, R.; Laganá, A. *Anal. Chim. Acta* 2007, *597*, 141-148.
[77] Berthiller, F.; Schuhmacher, R.; Buttinger, G.; Krska, R. *J. Chromatogr. A* 2005, *1062*, 209-216.
[78] Spanjer, M. C.; Rensen, P. M.; Scholten, J. M. *Food Addit. Contam.* 2008, *25*, 472-489.
[79] Lobeau, M.; De Saeger, S.; Sibanda, L.; Barna-Vetró, I.; Van Peteghem, C. *Anal. Chim. Acta* 2005, *538*, 57-61.
[80] Mortensen, G. K.; Strobel, B. W.; Hansen, H. C. B. *Anal. Bional. Chem.* 2003, *76*, 98-101.
[81] Delmulle, B.; De Saeger, S.; Adams, A.; De Kimpe, N.; Van Peteghem, C. *Rapid Commun. Mass Spectrom.* 2006, *20*, 771-776.
[82] Ren, Y.; Zhang, Y.; Shao, S.; Cai, Z.; Feng, L.; Pan, H.; Wang, Z. *J. Chromatogr. A* 2007, *1143*, 48-64.
[83] Sulyok, M.; Berthiller, F.; Krska, R.; Schuhmacher, R. *Rapid Commun. Mass Spectrom.* 2006, *20*, 2649-2659.
[84] Wang, J.; Zhou, Y.; Wang, Q. *Food Chem.* 2008, *107*, 970-976.
[85] Muscarella, M.; Lo Magro, S.; Nardiello, D.; Palermo, C.; Centonze, D. *J. Chromatogr. A* 2008, *1203*, 88-93.
[86] Tanaka, H.; Takino, M.; Sugita-Konishi, Y.; Tanaka, T. *Rapid Commun. Mass Spectrom.* 2006, *20*, 1422-1428.
[87] Rosenberg, E.; Krska, R.; Wissiack, R.; Kmetov, V.; Josephs, R.; Razzazi, E.; Grasserbauer, M. *J. Chromatogr. A* 1998, *819*, 277-288.
[88] De Saeger, S.; Dumoulin, F.; Van Peteghem, C. *Rapid Commun. Mass Spectrom.* 2004, *18*, 2661-2668.
[89] Cavaliere, C.; Foglia, P.; Pastorini, E.; Samperi, R.; Laganá, A. *Rapid Commun. Mass Spectrom.* 2005, *19*, 2085-2093.
[90] Goryacheva, I. Y.; De Saeger, S.; Delmulle, B.; Lobeau, M.; Eremin, S. A.; Barna-Vetró, I.; Van Peteghem, C. *Anal. Chim. Acta* 2007, *590*, 118-124.
[91] Jelen, H. H.; Wasowicz, E. *J. Chromatogr. A* 2008, *1215*, 203-207.
[92] Sulyok, M.; Krska, R.; Schuhmacher, R. *Anal. Bional. Chem.* 2007, *389*, 1505-1523.
[93] Garrido-Frenich, A.; Martínez-Vidal, J. L.; Romero-González, R.; Aguilera-Luiz, M. M. *Food Chem.* 2009, *117*, 705-712.
[94] Goryacheva, I. Y.; de Saeger, S.; Eremin, S. A.; van Peteghem, C. *Food Addit. Contam.* 2007, *24*, 1169-1183.
[95] Schothorst, R. C.; Jekel, A. A. *Food Chem.* 2001, *73*, 111-117.
[96] Krska, R.; Molinelli, A. *Anal. Bional. Chem.* 2007, *387*, 145-148.

[97] Songsermsakul, P.; Razzazi-Fazeli, E. *J. Liq. Chromatogr. Relat. Technol.* 2008, *31*, 1641-1686.
[98] Shephard, G. S.; Sewram, V. *Food Addit. Contam.* 2004, *21*, 498-505.
[99] Yagen, B.; Sintov, A.; Bialer, M. *J. Chromatogr.* 1986, *356*, 195-201.
[100] Sokolovic, M.; Simpraga, B. *Food Control* 2006, *17*, 733-740.
[101] Stroka, J.; van Otterdijk, R.; Anklam, E. *J. Chromatogr. A* 2000, *904*, 251-256.
[102] Caldas, E. D.; Silva, A. C. S. *J. Agric. Food Chem.* 2007, *55*, 7974-7980.
[103] Braicu, C.; Puia, C.; Bodoki, E.; Socaciu, C. *J. Food Qual.* 2008, *31*, 108-120.
[104] Krska, R.; Baumgartner, S.; Josephs, R. *Fresenius J. Anal. Chem.* 2001, *371*, 285-299.
[105] Krska, R.; Josephs, R. *Anal. Bional. Chem.* 2001, *369*, 469-476.
[106] Tanaka, T.; Yoneda, A.; Inoue, S.; Sugiura, Y.; Ueno, Y. *J. Chromatogr. A* 2000, *882*, 23-28.
[107] Nielsen, K. F.; Thrane, U. *J. Chromatogr. A* 2001, *929*, 75-87.
[108] Onji, Y.; Auki, Y.; Tani, N.; Umebayashi, K.; Kitada, Y.; Dohi, Y. *J. Chromatogr. A* 1998, *815*, 59-65.
[109] Valle-Algarra, F.; Medina, A.; Gimeno-Adelantado, J. V.; Llorens, A.; Jiménez, M.; Mateo, R. *Talanta* 2005, *66*, 194-201.
[110] Soleas, G. J.; Yan, J.; Goldberg, D. M. *J. Agric. Food Chem.* 2001, *49*, 2733-2740.
[111] Tabata, S.; Iida, K.; Suzuki, J.; Kimura, K.; Ibe, A.; Saito, K. *J. Food Hyg. Soc. Jpn.* 2004, *45*, 245-249.
[112] Josephs, R. D.; Krska, R.; Grasserbauer, M.; Broekaert, J. A. C. *J. Chromatogr. A* 1998, *795*, 297-304.
[113] Kotal, F.; Holadova, K.; Hajslova, Poustka, J.; Radova, Z. *J. Chromatogr. A* 1999, *830*, 219-225.
[114] Olsson, J.; Börjesson, T.; Lundstedt, T.; Schürer, J. *Int. J. Food Microbiol.* 2002, *72*, 203-214.
[115] Herzallah, S. M. *Food Chem.* 2009, *114*, 1141-1146.
[116] Aresta, A.; Cioffi, N.; Palmisano, F.; Zambonin, C. G. *J. Agric. Food Chem.* 2003, *51*, 5232-5237.
[117] Akiyama, H.; Goda, Y.; Tanaka, T.; Toyoda, M. *J. Chromatogr. A* 2001, *932*, 153-157.
[118] Goncalez, E.; Nogueira, J. H. C.; Fonseca, H.; Felicio, J. D.; Pino, F. A.; Correa, B. *Int. J. Food Microbiol.* 2008, *123*, 184-190.
[119] Yang, M. H.; Chen, J. M.; Zhang, X. H. *Chromatographia* 2005, *62*, 499-504.

[120] Tavcar-Kalcher, G.; Vrtac, K.; Pestevsek, U.; Vengust, A.; *Food Control* 2007, *18*, 333-337.
[121] O'Riordan, M. J.; Wilkinson, M. G. Food Control 2009, 20, 700-705.
[122] Khoury, A.; Rizk, T.; Lteif, R.; Azouri, H.; Delia, M. L.; Lebrihi, A. *Food Chem. Toxicol.* 2008, *46*, 2244-2250.
[123] Silva, L.; Fernández-Franzón, M.; Font, G.; Pena, A.; Silveira, I.; Lino, C.; Mañés, J. *Food Chem.* 2009, *112*, 1031-1037.
[124] Neuhof, T.; Koch, M.; Rasenko, T.; Nehls, I. *J. Agric. Food Chem.* 2008, *56*, 7566-7571.
[125] Jiménez, M.; Mateo, J. J.; Mateo, R. *J. Chromatogr. A* 2000, *870*, 473-481.
[126] Dall'Asta, C.; Galaverna, G.; Biancardi, A.; Gasparini, M.; Sforza, S.; Dossena, A.; Marchelli, R. *J. Chromatogr. A* 2004, *1047*, 241-247.
[127] Mateo, J. J.; Mateo, R.; Hinojo, M. J.; Llorens, A.; Jiménez, M. *J. Chromatogr. A* 2002, *955*, 245-256.
[128] Gaspar, E. M. S. M.; Lucena, A. F. F. *Food Chem.* 2009, *114*, 1576-1582.
[129] Shephard, G. S.; Leggott, N. L. *J. Chromatogr. A* 2000, 882, 17-22.
[130] Boonzaaijer, G.; Bobeldijk, I.; van Osenbruggen, W. A. *Food Control* 2005, *16*, 587-591.
[131] Katerere, D. R.; Stockenström, S.; Shephard, G. S. *Food Control* 2008, *19*, 389-392.
[132] Iha, M. H.; Souza, S. V. C.; Sabino, M. *Food Control* 2009, *20*, 569-574.
[133] Nguyen, M. T.; Tozlovanu, M.; Tran, T. L.; Pfhol-Leszkowicz, A. *Food Chem.* 2007, *105*, 42-47.
[134] Shephard, G. S.; Fabiani, A.; Stockenstrom, S.; Mshicilelli, N.; Sewram, V. *J. Agric. Food Chem.* 2003, *51*, 1102-1106.
[135] Zöllner, P.; Jodlbauer, J.; Lindner, W. *J. Chromatogr. A* 1999, *858*, 167-174.
[136] [136] Berthiller, F.; Dall'Asta, C.; Schuhmacher, R.; Lemmens, M.; Adam, G.; Krska, R. *J. Agric. Food Chem.* 2005, *53*, 3421-3425.
[137] Nielsen, K. F.; Smedsgaard, J. *J. Chromatogr. A* 2003, *1002*, 111-136.
[138] Ventura, M.; Gómez, A.; Anaya, I.; Díaz, J.; Broto, F.; Agut, M.; Comellas, L. *J. Chromatogr. A* 2004, *1048*, 25-29.
[139] Zollner, P.; Jodlbauer, J.; Lindner, W. *J. Chromatogr. A* 1999, *858*, 167-174.
[140] Rychlik, M.; Schieberle, P. *J. Agric. Food Chem.* 1999, *47*, 3749-3755.
[141] Kataoka, H.; Itano, M.; Ishizaki, A.; Saito, K. *J. Chromatogr. A* 2009, *1216*, 3746-3750.
[142] Sewram, V.; Nair, J. J., Nieuwoudt, W.; Leggott, N. L., Shephard, G. S. *J. Chromatogr. A*, 2000, *897*, 365-374.

[143] Takino, M.; Daishima, S.; Nakahara, T. *Rapid Commun. Mass Spectrom.* 2003, *17*, 1965-1972.
[144] Josephs, J. L. *Rapid Commun. Mass Spectrom.* 1996, *10*, 1333-1344.
[145] Newkirk, D. K.; Benson, R. W.; Howard, P. C.; Churchwell, M. I.; Doerge, D. R.; Roberts, D. W. *J. Agric. Food Chem.* 1998, *46*, 1677-1688.
[146] Cirillo, T.; Ritieni, A.; Visone, M.; Cocchieri, R. A. *J. Agric. Food Chem.* 2003, *51*, 8128-8131.
[147] [147] Berger, U.; Oehme, M.; Kuhn, F. *J. Agric. Food Chem.* 1999, *47*, 4240-4245.
[148] Razzazi-Fazeli, E.; Rabus, B.; Cecon, B.; Böhm, J. *J. Chromatogr. A* 2002, *968*, 129-142.
[149] Klötzel, M.; Gutsche, B.; Lauber, U.; Humpf, H. U. *J. Agric. Food Chem.* 2005, *53*, 8904-8910.
[150] Sulyok, M.; Krska, R.; Schuhmacher, R. *Food Addit. Contam.* 2007, *24*, 1184-1195.

INDEX

A

absorption, 42
acceptor, 18
accessibility, 30
accuracy, 31, 67
acetate, 20, 30, 32, 33, 45, 47, 48, 54, 56, 61, 62, 63, 64, 65
acetic acid, 19, 24, 26, 40, 45, 46, 59, 65, 66
acetone, 20, 21, 27, 30, 32, 33
acetonitrile, 17, 20, 23, 24, 40, 41, 43, 56
acidic, 23, 40
additives, 2, 71
adsorption, 15
aflatoxins, 1, 5, 10, 11, 18, 23, 25, 30, 31, 40, 41, 43, 53, 64, 68
agent, 34, 41
agricultural, ix, 1, 2, 30, 43
agricultural commodities, 1, 2
alkaline, 23
alternative, 17, 23, 41, 67
alternatives, 13, 17
amine, 9
amino, 23
ammonia, 40, 63
ammonium, 47, 54, 56, 61, 62, 63, 64, 65
animals, 1, 2, 8
application, ix, x, 14, 17, 24, 43, 68
ascorbic acid, 27

atmospheric pressure, 53
availability, 12

B

beer, 2, 5, 8, 14, 19
beverages, 8
binding, 8
biosynthesis, 68
blood, 8
body weight, 11, 12
breakfast, 10, 23
bromine, 41, 47
buffer, 40, 47, 49, 54

C

calibration, 56, 67
cancer, 9
carbon, 9, 23
carboxylic, 40, 53, 54
carboxylic groups, 54
carcinogen, 5, 8
carcinogenic, 8
casein, 8
cell, 41

cereals, 2, 5, 8, 10, 11, 22, 23
cheese, 41, 47
chloride, 32, 42, 50, 51
chloroform, 14, 22, 30, 33
chromatograms, 57, 59
chromatographic technique, x, 3, 13, 24
chromatography, ix, 2, 35, 42, 54
classes, 23, 42, 43
cleanup, ix, 35
coffee, 2, 5, 8, 11, 23, 27
commodity, 43
complexity, 3, 29
compliance, 2
components, 20, 42
composition, 15, 41, 67
concentration, 6, 10, 12, 13, 17, 35, 53
confidence, 53, 67
conservation, 5
consumption, 8, 9
contaminants, ix, 2, 9, 13, 69, 71
contamination, ix, 1, 8, 10, 12
control, ix, 12
corn, 9, 10
costs, 30
cows, 8
CRM, 36
cultivation, 5

D

dairy, 8
dairy products, 8
deconvolution, 35
degradation, 36
delivery, 13
densitometry, 31, 32
deoxynivalenol, 1, 9, 10, 11, 14, 23, 34, 36, 38, 54, 55, 63
derivatives, 34, 42
detection techniques, 3, 41
dietary, 2, 11
dispersion, 22, 28
diversity, 2
drugs, 1

E

education, 68
electron, ix, 3
emission, 16, 40, 42
environmental conditions, 5
environmental factors, 10
enzyme-linked immunosorbent assay, 29
enzymes, 10
epoxy, 9
ethanol, 14, 52
ethyl acetate, 20, 30, 32, 33, 48
excitation, 16, 40, 42
exposure, 8, 10
extraction, 3, 13, 14, 15, 16, 17, 18, 19, 22, 23, 24, 28, 43

F

false negative, 43
false positive, 34, 43
family, 8
FAO, 11, 69, 71
fat, 17
FID, 29, 34, 36, 38, 39
filtration, 16
financial support, 68
flame, 29
flight, x, 35, 39, 66
fluorescence, ix, 3, 31, 33, 40, 41, 42, 43
fluorine, 34
food additives, 2, 71
food commodities, 5, 8
food products, 33
food safety, 2, 29
foodstuffs, 2, 5, 13, 22, 45, 59, 67, 71
fruit juice, 56
fruit juices, 56
funding, 68
fungal, 5, 10, 43, 66
fungal infection, 10
fungal metabolite, 66
fungi, ix, x, 1, 5, 68
fusarium, 1, 8, 9, 10, 24, 71

Index

G

gas, ix, 2, 35
gas chromatograph, ix, 2, 35
gel, 30, 31, 32, 33
genetic mutations, 10
glucose, 14
glycol, 14, 22, 28
grain, 10, 25
grains, 22, 23, 26
grapes, 6, 26
groups, 9, 14, 54

H

handling, 15
health, 1, 2, 68
heterogeneity, 13
hexane, 17, 20, 21, 23
high pressure, 57
high resolution, 68
HPLC, 56
human, ix, 2, 8
humans, 1, 2, 5, 8
hydrogen, 14
hydroxyl, 9, 14
hydroxyl groups, 14

I

identification, 2, 30, 33, 67, 68
identity, 35
immunoassays, 2
immunosuppressive, 5, 10
incidence, 9
infection, 9, 10
injection, 3, 33
instruments, 57, 67
interference, 14
International Agency for Research on Cancer, 5
International Organization for Standardization, 30
iodine, 41, 48

ion exchangers, 17
ionization, 29, 39, 53, 54, 55, 56, 59, 60, 61, 62, 63, 64, 65, 66
ions, 34, 35, 54, 67
isolation, 14
isotope, 56

L

lactones, 64
laser, 52
legislation, 2, 11, 29
light scattering, 41
limitations, 43
lipids, 23
liquid chromatography, ix, 2, 56
liquid phase, 14
liver, 8
liver cancer, 8
losses, 24

M

maize, 8, 9, 11, 18, 26, 55, 57, 59, 71
manipulation, 41
mass spectrometry, x, 3, 56, 68
matrix, 11, 12, 13, 14, 18, 23, 29, 42, 53, 56, 67, 68
membranes, 9
metabolite, 8
metabolites, ix, x, 1, 66
meteorological, 6
milk, 2, 8, 10, 11, 14, 17, 21, 41, 47
millet, 10
mitochondrial, 10
modules, 57
moisture, 10, 34
moisture content, 10
molecular mass, 42
molecular structure, 9, 54
molecular weight, 67
molecules, 40, 56
monoclonal, 2
mouse, 23

mutations, 10

N

NaCl, 15, 16, 26
naphthalene, 41
natural, ix, 1, 9, 40
nephrotoxic, 5
nitric acid, 41
normal, 41, 42
nuts, 8, 11, 22

O

ODS, 45, 47, 48, 52, 59, 62
oil, 18, 21, 22
olive, 18
organic, 14, 17, 20, 27, 56
organic solvent, 14, 17, 56
organic solvents, 14
oxidation, 27

P

pasta, 9
PDMS, 18, 22
pesticide, 2
pesticides, 1
phosphate, 18, 19, 27, 49
photoionization, 53, 66
physicochemical, 13
physicochemical properties, 13
pigments, 23
plasma membrane, 10
PLS, 20
polarity, 14, 22, 54, 55, 56
polydimethylsiloxane, 22
polyethylene, 14
poor, 56
population, 8, 13
potassium, 32, 41
PPI, 53, 66
pressure, 53, 57, 66
producers, 12

production, 6, 35
program, 68
protein, 8, 9
proteins, 17
protocols, 33, 40
PRP, 60
public, ix, 1, 3
public health, 1, 3
pulmonary edema, 9
purification, 13

Q

quadrupole, 35, 39, 53, 66, 67

R

range, 2, 10, 12, 23, 24, 33, 43, 67
raw material, 8
reagent, 34, 42
recovery, 17, 34, 41
red wine, 6, 15, 16
regulations, ix, 29, 56, 69
relationship, 6
reliability, 43, 56
residues, 2
resolution, 35, 53, 56, 67, 68
retention, 43
rice, 10
risk, 2
risks, 2
RNA, 9
roasted coffee, 23
robotic, 15
rural, 9
rural areas, 10

S

safety, 1
saline, 18, 19
sampling, 13
scattering, 41, 52
selecting, 53, 56

selectivity, 3, 42, 44
sensitivity, x, 3, 17, 29, 34, 41, 42, 53, 54, 55, 56, 67, 68
separation, ix, 3, 30, 31, 35, 43, 57
silica, 31, 38
sodium, 45, 53
soil, 10
solid phase, 14, 18, 22, 28
solvent, 3, 22, 24, 32
solvents, 22, 43
sorbents, 14, 17, 23
species, ix, 1, 9, 10, 43, 54
specificity, 14
spectrum, 42
speed, 24, 29, 56
stability, 34, 36, 41
stabilize, 41
standards, x, 56, 67
storage, 1, 68
strain, 5
strategies, 56
styrene, 17, 20, 60
substances, 43
surface area, 22
switching, 54, 56
synthesis, 9
systems, 13

T

tandem mass spectrometry, 3
temperature, 10, 45, 49, 59
thermal stability, 36
time, x, 14, 18, 24, 25, 26, 27, 32, 34, 35, 40, 41, 43, 56
time consuming, 34
tolerance, 11

toluene, 14, 30, 33
toxic, 1, 2, 5, 9
toxicity, ix, 8
toxin, 2, 9, 10, 15, 36, 38, 42, 55, 63, 67
trade, 30
transport, 1
traps, 67
trifluoroacetic acid, 41
tumours, 10

U

United Nations, 11
urine, 8

V

visible, 33
voids, 27

W

water, 17, 19, 20, 21, 22, 24, 25, 26, 27, 30, 31, 32, 40, 41, 42, 52, 53
wavelengths, 42
wheat, 8, 9, 10, 24, 25, 26, 34
WHO, 11, 12, 69, 70, 71
wine, 2, 5, 8, 11, 14, 16, 17, 18, 19
World Health Organization, 11, 71

Y

yield, 40, 68